T0298346

Human Factors on the Flight Deck

A Practical Guide for Design, Modelling and Evaluation

This book presents the Human Factors methodologies and applications thereof that can be utilised across the design, modelling and evaluation stages of the design lifecycle of new technologies entering future commercial aircraft.

As advances are made to the architecture of commercial aircraft cockpits, *Human Factors on the Flight Deck* argues that it is vitally important that these new interfaces are safely incorporated and designed in a way that is usable to the pilot. Incorporation of Human Factors is essential to ensuring that engineering developments to avionic systems are integrated such that pilots can maintain safe interactions while gaining information of value. Case study examples of various technological advancements during their early conceptual stages are given throughout to highlight how the methods and processes can be applied across each stage.

The text will be useful for professionals, graduate students and academic researchers in the fields of aviation, Human Factors and ergonomics.

Transportation Human Factors: Aerospace, Aviation, Maritime, Rail, and Road Series
Series Editor: Professor Neville A. Stanton, University of Southampton, UK

Eco-Driving
From Strategies to Interfaces
Rich C. McIlroy and Neville A. Stanton

Driver Reactions to Automated Vehicles
A Practical Guide for Design and Evaluation
Alexander Eriksson and Neville A. Stanton

Systems Thinking in Practice
Applications of the Event Analysis of Systemic Teamwork Method
Paul Salmon, Neville A. Stanton, and Guy Walker

Individual Latent Error Detection (I-LED)
Making Systems Safer
Justin R.E. Saward and Neville A. Stanton

Driver Distraction
A Sociotechnical Systems Approach
Kate J. Parnell, Neville A. Stanton, and Katherine L. Plant

Designing Interaction and Interfaces for Automated Vehicles
User-Centred Ecological Design and Testing
Neville Stanton, Kirsten M.A. Revell, and Patrick Langdon

Human-Automation Interaction Design
Developing a Vehicle Automation Assistant
Jediah R. Clark, Neville A. Stanton, and Kirsten Revell

Assisted Eco-Driving
A Practical Guide to the Design and Testing of an Eco-Driving Assistance System (EDAS)
Craig K. Allison, James M. Fleming, Xingd A. Yan, Roberto Lot, and Neville A. Stanton

Safety Matters
An Introduction to Safety Science
David O'Hare

Human Factors on the Flight Deck
A Practical Guide for Design, Modelling and Evaluation
Katie J. Parnell, Victoria A. Banks, Rachael A. Wynne, Neville A. Stanton, and Katherine L. Plant

For more information about this series, please visit: https://www.crcpress.com/Transportation-Human-Factors/book-series/CRCTRNHUMFACAER

Human Factors on the Flight Deck

A Practical Guide for Design, Modelling and Evaluation

Katie J. Parnell, Victoria A. Banks,
Rachael A. Wynne, Neville A. Stanton
and Katherine L. Plant

CRC Press
Taylor & Francis Group
Boca Raton London New York

CRC Press is an imprint of the
Taylor & Francis Group, an **informa** business

Designed cover image: © GE Aerospace

First edition published 2023
by CRC Press
6000 Broken Sound Parkway NW, Suite 300, Boca Raton, FL 33487-2742

and by CRC Press
4 Park Square, Milton Park, Abingdon, Oxon, OX14 4RN

CRC Press is an imprint of Taylor & Francis Group, LLC

Library of Congress Cataloging-in-Publication Data
Names: Parnell, Katie J., author. | Banks, Victoria A., author. | Wynne, Rachael A., author. |
Stanton, Neville A. (Neville Anthony), 1960– author. | Plant, Katherine L., author.
Title: Human factors on the flight deck : a practical guide for design, modelling and evaluation /
Katie J. Parnell, Victoria A. Banks, Rachael A. Wynne, Neville A. Stanton and Katherine L. Plant.
Other titles: Transportation human factors (CRC Press)
Description: First edition. | Boca Raton, FL : CRC Press/Taylor & Francis Group, LLC, 2023. |
Series: Transportation human factors: aerospace, aviation, maritime, rail, and road series |
Includes bibliographical references and index. |
Identifiers: LCCN 2022051456 (print) | LCCN 2022051457 (ebook) |
ISBN 9780367754471 (hardcover) | ISBN 9781032470702 (paperback) |
ISBN 9781003384465 (ebook)
Subjects: LCSH: Airplanes–Cockpits–Design and construction. |
Airplanes–Design and construction–Human factors. |
Human engineering. | Airplanes–Piloting–Human factors.
Classification: LCC TL681.C6 P37 2023 (print) | LCC TL681.C6 (ebook) |
DDC 629.134/1–dc23/eng/20221122
LC record available at https://lccn.loc.gov/2022051456
LC ebook record available at https://lccn.loc.gov/2022051457

ISBN: 9780367754471 (hbk)
ISBN: 9781032470702 (pbk)
ISBN: 9781003384465 (ebk)

DOI: 10.1201/9781003384465

Typeset in Times
by Newgen Publishing UK

Contents

Preface

This book presents the methods, measures and processes for applying a user-centred approach to the design and integration of future flight deck technology. It makes the case for including end-users throughout all stages of the design of new cockpit technologies and provides the tools and knowledge to achieve this. User-centred design places the needs, motivations and constraints at the centre of the design requirements to maximise safety and usability. The focus of this book is on the development of new cockpit technologies within commercial aircraft and therefore the user base is commercial airline pilots. The work presented within this book has been conducted alongside aerospace manufactures, systems engineers, test pilots, airline pilots and Human Factors practitioners.

The book provides an overview of user-centred design and the methods, literature and research practices that surround it. It then focuses on three key stages of the design process: designing, modelling and evaluating. These key stages are integrated into the user-centred design process, upon which the book is structured. Real-world case studies are presented throughout the book to demonstrate the application of user-centred principles and methods within each of these different design stages.

This book will be of interest to Human Factors practitioners who are interested in understanding how a range of different Human Factors methods can be applied across the design life cycle, as well as how to apply the methods. Aviation engineers will be interested to understand how the Human Factors method can be utilised to complement the design of aerospace technologies, and the case studies will bring this to life. More generally, the book is also of interest to researchers and students who are interested in understanding how research best practice in the aviation domain is conducted, as well as the benefits and challenges of conducting user-centred research.

Acknowledgements

We would like to thank the partners we worked with during the Innovate UK-funded Open Flight Deck project: Rolls Royce, GE Aviation Systems, BAE Systems and Coventry University. We would like to thank the various participants who contributed to the studies discussed in this book. We would also like to thank the subject matter experts who contributed to the research activities in this book, in particular, Pete Beecroft and Tom Griffin.

This publication is, in part, based on work performed in the Open Flight Deck project, which received funding from the ATI Programme, a joint government and industry investment to maintain and grow the UK's competitive position in civil aerospace design and manufacture. The programme, delivered through a partnership between the Aerospace Technology Institute (ATI), Department for Business, Energy & Industrial Strategy (BEIS) and Innovate UK, addressed technology, capability and supply-chain challenges. The front cover shows the Open Flight Deck simulator developed by GE Aerospace under the Open Flight Deck programme.

About the Authors

Katie J. Parnell, BSc, EngD, C.ErgHF, is a chartered Human Factors practitioner and a Senior Research Fellow in Human Factors Engineering at the University of Southampton. She completed her engineering doctorate in 2018 and published the outputs of her work in the CRC Press book '*Driver Distraction: A Sociotechnical Systems Approach*'. Dr Parnell has extensive experience in applying Human Factors methods and theories across numerous disciplines including aviation, road transport, automation and uncrewed aerial vehicles. Her research interests particularly centre around include user-centred design and the application of sociotechnical systems theory to assess the interaction between new technologies and the social systems that they function within. She has published numerous articles in the fields of Human–Computer Interaction, Safety Science, Accident Prevention and Ergonomics. In 2021, Katie was awarded the Saul Aviation Award by The Honourable Air Pilots Association for her contribution to aviation safety research. In 2022, Katie was awarded an Anniversary Fellowship from the University of Southampton to undertake research on the inclusive design of future transport systems.

Victoria A. Banks, BSc, EngD, C.ErgHF, is a Visiting Research Fellow in Human Factors Engineering at the University of Southampton. Her primary research interests include understanding operator decision-making in the context of hybrid autonomous teams as well as Human Factors integration into systems architecture. She has utilised her skills in modelling complex sociotechnical systems, predicting and analysing human performance and interface design in various domains including automotive, aerospace and maritime sectors. Her research portfolio includes published works in books, scientific journals and conference proceedings and, in 2020, Victoria was the recipient of the Honourable Company of Air Pilots (Saul) Prize for research into Aviation Safety. She continues to apply insights from the human sciences to understand and solve problems in both industry and academic projects that centre around the broad themes of 'human–machine teaming', 'hybrid autonomous systems', 'operator monitoring', 'human–system integration' and 'human–robot collaboration'.

Rachael A. Wynne, B. PsySci (Hons), PhD, is an Associate Lecturer in the School of Psychological Sciences at The University of Newcastle, Australia, and a Visiting Research Fellow in Human Factors Engineering at the University of Southampton, UK. Her primary research interests are in the field of cognitive psychology and Human Factors, studying how we visually engage with the world around us, and how our experiences and expectations guide our visual search and situation awareness. Rachael has applied this to the areas of driving (learner drivers, cyclists, distraction and hazard perception), aviation (how pilots engage with new technologies) and e-sports (comparing different aspects of playing experience). She is currently working on a project evaluating Human–Machine Interfaces in vehicles in the Australian market and their impact on driver distraction, as well as projects relating to false memory and inattentional blindness.

Neville A. Stanton, PhD, DSc, Professor Emeritus, is a Chartered Psychologist, Chartered Ergonomist and Chartered Engineer. His research interests include modelling, predicting, analysing and evaluating human performance in systems as well as designing the interfaces and interactions between humans and technology. Professor Stanton has worked on the design of automobiles, aircraft, ships and control rooms over the past 30 years, on a variety of automation projects. He has published 50 books and over 400 journal papers on Ergonomics and Human Factors. In 1998 he was presented with the Institution of Electrical Engineers Divisional Premium Award for research into System Safety. The Institute of Ergonomics and Human Factors in the UK awarded him The Otto Edholm Medal in 2001, The President's Medal in 2008 and 2018, The Sir Frederic Bartlett Medal in 2012 and The William Floyd Medal in 2019 for his contributions to basic and applied ergonomics research. The Royal Aeronautical Society awarded him and his colleagues the Hodgson Prize in 2006 for research on design-induced, flight-deck, error published in *The Aeronautical Journal*. The University of Southampton awarded him a Doctor of Science in 2014 for his sustained contribution to the development and validation of Human Factors methods.

Katherine L. Plant, PhD, C. ErgHF, Fellow of HEA, is an Associate Professor in Human Factors Engineering in the Transportation Research Group, Faculty of Engineering and Physical Sciences, University of Southampton. Katherine has worked extensively in the field of aeronautical critical decision-making; being awarded the Honourable Company of Air Pilots (Saul) Prize for research into Aviation Safety (2014) and publishing '*Distributed Cognition and Reality: How pilots and crews make decisions*' (CRC Press, 2016). Katherine is the Director of the Human Factors Engineering team, who, in 2018, were the recipients of The Chartered Institute of Ergonomics and Human Factors Presidents Award. Katherine manages an extensive research portfolio covering the fields of aviation, command and control, trust in autonomous systems and road safety.

Abbreviations

ACARS	Aircraft Communications Addressing and Reporting System
ATB	Air Turn Back
ATC	Air Traffic Control
ATPL	Airline Transport Pilot Licence
ATSB	Australian Transport Safety Bureau
BU	Bottom-Up
CDM	Critical Decision Method
CDU	Control Display Unit
CFMS	Connected Flight Management System
CLR	Clear
CONOPs	Concept of Operations
CPL	Commercial Pilot License
DODAR	Diagnose, Options, Decision, Assign task, Review
DV	Dependent Variable
DwI	Design with Intent
EAST	Event Analysis of Systemic Teamwork
ECAM	Electronic Centralised Aircraft Monitor
EFB	Electronic Flight Bag
EICAS	Engine Indicating and Crew Alerting System
EMA	Engine Monitoring Assistant
ERGO	Ethical Research Governance Office
ESD	Event Sequence Diagram
FAA	Federal Aviation Administration
FMC	Flight Management Computer
FMS	Flight Management System
FSS	Future Systems Simulator
HCP	Human-Centred Processes
HEI	Human Error Identification
HET	Human Error Template
HF	Human Factors
HMI	Human–Machine Interface
HTA	Hierarchical Task Analysis
ISO	International Organisation for Standardisation
ITCZ	Intertropical Convergence Zone
IV	Independent Variable
IVIS	In-Vehicle Information Systems
KPI	Key Performance Indicators
LCD	Liquid Crystal Display
MFD	Multi-Function Displays
NASA-TLX	National Aeronautics and Space Administration – Task Load Index
NDM	Naturalistic Decision-Making
NTSB	National Transportation Safety Board

OESD	Operational Event Sequence Diagram
OSA	Oil Starvation Avoidance
OSD	Operational Sequence Diagrams
PCAP	Projected Capacitive
PCM	Perceptual Cycle Model
PF	Pilot Flying
PIS	Participant Information Sheet
PM	Pilot Monitoring
PNF	Pilot Not Flying
QRH	Quick Reference Handbook
RPDM	Recognition Primed Decision Model
SA	Situation Awareness
SAR	Situation Assessment Record
SAW	Schema, Action, World
SD	Systems Display
SHERPA	Systematic Human Error Reduction and Prediction Approach
SME	Subject Matter Expert
SUS	System Usability Scale
SWARM	Schema World Action Research Method
TAFEI	Task Analysis for Error Identification
TD	Top-Down
T-DODAR	Time, Diagnose, Options, Decision, Assign task, Review
UCD	User-Centred Design
UTC	Universal Time Coordinated
VDT	Visual Display Terminals
WUCI	Warm-up/Cool-down indicator

1 Introduction to Human Factors in Aviation

1.1 INTRODUCTION

This chapter provides an overview of the application of Human Factors to the aviation domain. It describes the domain as a system of systems, with numerous complexly interacting actors, elements and regulations that work together to ensure safe and efficient performance. It then introduces the micro, meso and macro levels of systems analysis that are prevalent within the study of Human Factors. An explanation of each of these levels, with respect to the aviation domain in provided. Examples of adverse events and incidents that occurred due to poor design and a lack of awareness for the end-user are presented to highlight the motivation for applying Human Factors to the aviation domain. This sets the scene for the need to incorporate user-centred design methods and principles within the design of future flight deck systems. The chapter concludes with an overview of the structure of the book.

1.2 HUMAN FACTORS IN AVIATION

Human Factors in Aviation is essential for the safety and efficiency of commercial airlines, passenger, cargo and military operations, and for the well-being of their employees and passengers. These factors extend beyond the individual aircraft to air traffic control and management, maintenance, regulatory bodies and policy makers. Human Factors has a long history of innovations in theory, methodology, science and application (Stanton et al, 2019a). For example, approaches have evolved from the examination of the activities of individual pilots, through to crew resource management to considering entire aviation systems and their emergent properties. Similarly, ergonomics methodologies have moved from focussing on individual tasks to entire systems, the constraints shaping behaviour and the culture of organisations.

Harris and Stanton (2010) argued that aviation is a system of systems. Some of this complexity is characterised in Figure 1.1, which shows that the aviation sociotechnical system comprises airports, aircraft, airlines, air traffic management and air traffic control together with all of the related subsystems (Stanton et al, 2019a). All of these systems interact with each other within the rules and regulations of the aviation authorities around the world.

DOI: 10.1201/9781003384465-1   1

SocioTechnical Systems

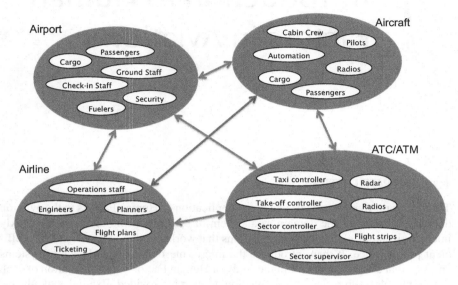

FIGURE 1.1 Examples of some of the factors involved in aviation as a system of systems.

The systems paradigm is gaining ground in Human Factors (Salmon et al, 2017; Stanton and Harvey, 2017; Walker et al, 2017). Indeed, the importance of the systems approach has been recognised in aviation (Harris and Stanton, 2010; Stanton et al, 2019a). Within aviation systems there are distinct operational interdependencies (aircraft operations; maintenance; air traffic management/control) and each of these aspects has managerial independence (they may be run by private companies or public providers). Nevertheless, they are all bound by a set of common operating principles and international regulations for design, maintenance and operation. The inherent complexity of these operations is difficult to capture in their entirety, as they are distributed both in time and space.

1.3 MICRO, MESO AND MACRO LEVELS OF ANALYSIS

Levels of analysis in Human Factors follows the micro, meso and macro models prevalent in sociotechnical systems research (Grote et al, 2014). At the micro level, models of system performance focus on the immediate interactions between individuals and technology (what has traditionally been called Human Machine Interaction or Human Computer Interaction). At the meso-level, models of systems focus on collaboration between humans and between humans and technology in larger work units (what has traditionally been called teamwork). At the macro level, the focus is on the complexity of multi-layered systems (what has traditionally been called organisational analysis).

1.3.1 MICRO LEVEL

At the micro level, Chapanis (1999) recalls his work at the Aero Medical Laboratory in the early 1940s where he was investigating the problem of pilots and co-pilots retracting the landing gear instead of the landing flaps after landing. His investigations in the B-17 (known as the 'Flying Fortress') revealed that the toggle switches for the landing gear and the landing flaps were both identical and next to each other. Chapanis's insight into human performance enabled him to understand how the pilot might have confused the two toggle switches, particularly after the stresses of a combat mission. He proposed coding solutions to the problem: separate the switches (spatial coding) and/or shape the switches to represent the part they control (shape coding), so the landing flap switch resembles a 'flap' and the landing gear switch resembles a 'wheel'. Thus the pilot can tell by looking at, or touching, the switch what function it controls. In his book, Chapanis (1999) also proposed that the landing gear switch could be deactivated if sensors on the landing struts detect the weight of the aircraft.

Similarly, Grether (1949) reports on the difficulties of reading the traditional three-needle altimeter which displays the height of the aircraft in three ranges: the longest needle indicates 100s of feet, the broad pointer indicates 1000s of feet and the small pointer indicates 10,000s of feet. Previous work had shown that pilots frequently misread the altimeter. This error had been attributed to numerous fatal and non-fatal accidents. Grether devised an experiment to see if different designs of altimeter could have an effect on the interpretation time and the error rate. If misreading altimeters really was a case of 'designer error' rather than 'pilot error' then different designs should reveal different error rates. Grether tested six different variations of the dial and needle altimeter containing combinations of three, two and one needles with and without an inset counter as well as three types of vertically moving scale (similar to a digital display). Pilots were asked to record the altimeter reading. The results of the experiment showed that there were marked differences in the error rates for the different designs of the altimeters. The data also show that those displays that took longer to interpret also produced more errors. The traditional three-needle altimeter took some 7 seconds to interpret and produced over 11 percent errors of 1,000 feet or more. By way of contrast, the vertically moving scale altimeters took less than 2 seconds to interpret and produced less than 1 percent errors of 1,000 feet or more.

Both of these examples, one from control design and one from display design, suggest that it is not 'pilot error' that causes accidents; rather it is 'designer error', i.e., poor representation of system information output to the pilot and confusing system input devices. This notion of putting the blame on the last person in the accident chain (e.g., the pilot), has lost credibility in modern aviation research. Modern day researchers take a systems view of error (Read et al, 2021), by understanding the relationships between all the moving parts in a system, both human and technical, from concept, to design, to manufacture, to operation and maintenance (including system mid-life upgrades) and finally to dismantling and disposal of the system. What is new here is the assertion that design-induced errors may be predicted in advance of the aircraft becoming operational (Stanton and Baber, 1996, 2002).

1.3.2 MESO LEVEL

At the meso level, two infamous examples of aircraft crashes show that the way the crew interact with the flight deck systems can give rise to problems. The first has become known as the Kegworth accident in 1989 (Plant and Stanton, 2012) and the second as the second as the Air France 447 crash (AF447) in 2009 (Salmon et al, 2016). Plant and Stanton (2012) describe a British Midland Boeing 737–400 aircraft leaving Heathrow at 19:52 to return to Belfast with eight crew and 118 passengers. As the aircraft was climbing through 28,300 ft the outer panel of a blade in the fan of the No. 1 (left) engine detached. This gave rise to a series of compressor stalls in the No. 1 engine, which resulted in the airframe shuddering, smoke and fumes entering the cabin and flight deck and fluctuations of the No. 1 engine parameters. The crew made the decision to divert to East Midlands Airport. Believing that the No. 2 (right) engine had suffered damage, the crew throttled it back. The shuddering ceased as soon as the No. 2 engine was throttled back, which persuaded the crew that they had correctly dealt with the emergency, so they continued to shut it down. The No. 1 engine operated apparently normally after the initial period of vibration, during the subsequent descent however the No. 1 engine failed, causing the aircraft to strike the ground 2 nm from the runway. The ground impact occurred on the embankment of the M1 motorway. Forty-seven passengers died and 74 of the remaining 79 passengers and crew suffered serious injury. (The synopsis was adapted from the official Air Accident Investigation Branch report, AAIB, 1990). From the summary presented it is evident that the event was entangled with a host of issues; there was confusion on the flight deck over which engine was damaged, the crew were unable to elicit the correct information from their instruments, nor did any of the passengers or cabin crew question the flight deck over their cabin address which stated the right engine was faulty even though smoke had been seen from the left. Despite the Air Accident Investigation Branch blaming 'pilot error' for the crash (as is the case in many aircraft crashes), their report identified 31 safety recommendations, including: Equipment Design (13), Training (7), Standards and Certification (4), Maintenance (3), Flight Data Recorder (2), Air Traffic Control (1), Research and Development (1), Flight Manual (1) and Inspection (1). These 31 recommendations suggest that 'pilot error' played a minor (if any) part in the crash.

Salmon et al (2016) describe the Air France 447 incident that occurred on 31 May 2009 during a scheduled passenger flight from Rio de Janeiro, Brazil, to Paris, France. Just over three and a half hours after departure, at approximately 2.02 am Universal Time Coordinated (UTC), the Captain left the flight deck to take a scheduled rest break. Shortly afterwards, at 2.03.44 am, the Pilot Flying (PF) noted that the plane had entered the Intertropical Convergence Zone (ITCZ), which is an area close to the equator that experiences severe weather consistently. The PF subsequently called through to a flight attendant to warn of the impending turbulence and the need to take care. The aircraft's anti-icing system was then turned on. Upon entry into the ITCZ, the aircraft's pitot tubes froze due to the low air temperature. Shortly before 2.10 am, an alert sounded in the cockpit to notify the pilots that the autopilot was disconnecting. At 2.10.06 am, the PF remarked, 'I have the controls' and was acknowledged by the Pilot Not Flying (PNF). Following this, the PF put the aeroplane into a steep climb by pulling back on his sidestick, triggering a stall warning which subsequently sounded 75 times for the remainder of the flight. The plane gained altitude rapidly but lost speed quickly. The PF continued to apply nose up inputs with the PNF apparently unaware of this. Eventually, the aircraft went into a stall and began to lose altitude. After trying unsuccessfully to identify the problem and an appropriate procedure, the PNF called the Captain back into the cockpit. At 2.11.32 am, the PF announced 'I don't have control of the plane.' At 2.11.37 am, the PNF took control of the aeroplane. Six seconds later, the Captain returned to the cockpit and subsequently attempted to diagnose the situation. Both the PF and PNF informed the Captain that they had lost control of the aircraft and did not understand why. At 2.13.40 am, the PF told the Captain that he had 'had the stick back the whole time', at which point the PNF took control of the plane and applied nose down inputs in an attempt to prevent the stall and gain speed. Unfortunately, these actions were taken too late, and at 2.14.29 am, the voice recorder stopped as the plane crashed into the ocean.

In aviation, non-technical skills (i.e., those skills not related directly to control of the aircraft but are required to interact with other people and make decisions) are as important as the skills required to fly the aircraft. Crew Resource Management has been successful in the training of non-technical skills but has been criticised for not integrating this with the technical skills training. Typically, the training of non-technical skills is undertaken after the technical skills have been mastered. Arguably, these skills should be trained together as they are dependent on each other. Ideally, the training of both sets of skills would be integrated into the operational environment, but neither can make up for poor design of the flight deck.

1.3.3 Macro Level

Whist this book focuses on the micro and meso levels of analysis, for the sake of completeness of the introduction, the macro level is also included. At the macro level, Stanton et al (2019b) presented the findings of an observational study of cargo operations using the Event Analysis of Systemic Teamwork (EAST) method. The researchers recorded flight operations for three outbound and two inbound flights over a five-day period. From this they constructed task, social and information networks to represent four key phases of flight: (i) pre-flight checks and engines start, (ii) taxi and take-off, (iii) descent and landing, and (iv) taxi, park and shut-down. The networks present a detailed analysis and work audit of contemporary operations (see Stanton et al, 2019b). This analysis examined the work of those involved in flight operations from a variety of perspectives (e.g., dispatch, maintenance, loading, Air Traffic Control, Air Traffic Management and the flight deck) as shown in Figure 1.2. To some extent, this is an attempt to examine aviation as a system of systems as denoted in Figure 1.1 (although it was limited to the examination of a single cargo operator). Nevertheless, the study was a first-of-a-kind in breadth and scope of the evaluation.

Stanton et al (2019b) noted that contemporary operational procedures and processes have evolved over the decades to provide extremely safe ways of working. It was accepted that the observational studies were solely based on normal

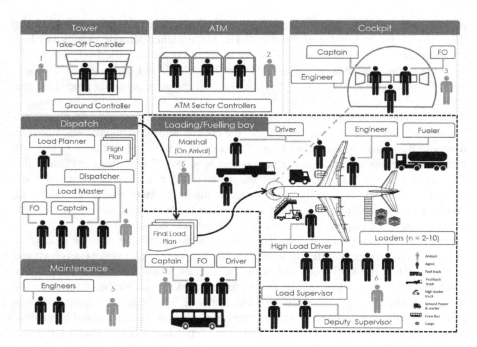

FIGURE 1.2 Rich picture of data collection from multiple perspectives in flight operations (with data collection analysts indicated by the grey figures numbered 1 to 6).

operations and no emergency events were observed that pushed it to the limits and its resilience tested. Thus, the insights related to current operations were limited to short-haul cargo operation for routine flights only. One of the main tenets of distributed cognition is that different operators each have their own unique view on, and contribution to, awareness of the system. This has been likened to the 'hive mind' for a technical system (Stanton et al, 2017a). It has been argued that the compatibility between these views is critical to support safe and efficient performance in systems. Conversely, incompatibilities in awareness between actors and/ or agents create threats to safety and performance (Sorensen and Stanton, 2015). In the observation of the system, no such threats were present. What the networks show is a considerable amount of interdependences between the tasks, agents and information. The degree of interconnectedness in the networks might indicate that the system would be resilient to disturbances (as aviation in inherently one of the safest forms of transportation).

1.4 HUMAN FACTORS AND AIRCRAFT DESIGN

At all levels of analysis in aviation systems, there is much that can be done to improve existing practices in Human Factors (Stanton et al, 2013, 2014). The selection of methods to evaluate design and evaluate systems should depend on at least six factors:

- Reliability and validity of methods;
- Criteria to be evaluated (e.g. speed, errors, satisfaction, usability, etc.);
- Acceptability and appropriateness of the methods;
- Abilities of the designers involved in the process;
- Access to end-users and subject matter experts; and
- Cost–benefit analysis of methods.

Establishing the validity of the methods makes good commercial sense, especially when one considers the costs associated with poor decisions, for example:

- Poor productivity of design personnel;
- Knock-on effects of difficult-to-use systems (e.g., speed and accuracy);
- Negative perception of systems throughout industry;
- Delays in certification of systems; and
- Cost of redesign.

We therefore argue that designers need to pay careful attention to all the stages in the method selection process, from developing appropriate selection criteria, through choosing methods and making the design decision, to validating the process as a whole.

Method selection is a closed-loop process with two feedback loops (as illustrated in Figure 1.3). The first feedback loop informs the selectors about the adequacy of the methods to meet the demands of the criteria, and the second feedback loop provides feedback about the adequacy of the device assessment process as a whole. The main stages in the process are identified as follows:

- Determine criteria: the criteria for assessment are identified;
- Compare methods against criteria: the methods are compared for their suitability;
- Apply methods: the methods are applied to the assessment of a device;
- Make decisions: the resultant data are evaluated and the device design is chosen; and
- Validate: the procedures for developing and assessing the criteria are validated.

Assessment criteria will be developed according to what information is required (e.g. speed of performance, error potential, user satisfaction and general aspects of device usability). Assessment methods may be determined by factors such as time available, access to end-users and cost of using the method. The use of Human Factors methods presented throughout this book was underpinned by the approach presented in Figure 1.3.

1.5   STRUCTURE OF THE BOOK

This book is written to provide an overview of the process for designing, modelling and evaluating future flight deck technological designs. Importantly, it

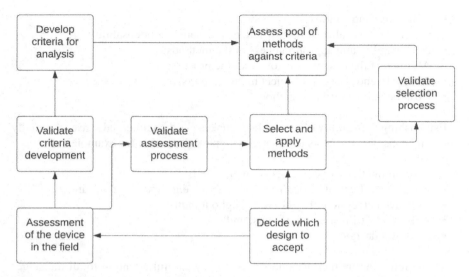

FIGURE 1.3 Human Factors methods selection process.

highlights the importance of including the end-user throughout this design process and provides methods and measures for how to achieve this. There are a number of case studies provided throughout the book that give real-world examples for how research and design should be conducted. To this end, the chapters of this book follow on from each other sequentially, building on each other to present the design process. The chapters can also be read individually should the reader be interested in specific aspects of the process. A brief overview of the contents of each chapter is provided below.

1.5.1 CHAPTER 1: INTRODUCTION

This first chapter provides an introduction to the work presented within this book and the importance of Human Factors within the aviation domain. It then presents an overview of the work throughout the following chapters of the book.

1.5.2 CHAPTER 2: HUMAN FACTORS APPROACH TO RESEARCH AND USER-CENTRED DESIGN

This chapter provides an overview of everything that needs to be considered when conducting experimental or research design when using human participants. Human Factors research is highly dependent on the involvement of human users within the research process and there are a number of important things that need to be considered from the outset when deciding to undertake this type of work. This is considered with a focus on the application to the aviation domain. This chapter will introduce you to the different types of experimental design that can be conducted as well as the data that be obtained. It also provides an introduction to user-centred design which is the key element that runs throughout this book.

1.5.3 Chapter 3: Human Factors Measures and Methods

This chapter presents the different measures and methods that are relevant to the application of Human Factors to the aviation domain. It highlights the important distinction between subjective and objective measures within the data collection process and how each should be applied. Methods are presented that are applied throughout the remaining chapters of the book; therefore this chapter is a useful reference to the later chapters where case study examples are provided. The benefits and limitations of the different methods are also considered.

1.5.4 Chapter 4: Defining Usability Criteria

The purpose of applying Human Factors principles to the design of new flightdeck technologies is to ensure that they are usable and safe. This chapter provides an overview of what it means for designs to be usable by outlining a set of usability criteria this are specific to the aircraft cockpit. This includes a summary of relevant literature in the field. Importantly, this discussion of usability discusses the importance of 'context of use' in assessing and defining usability.

1.5.5 Chapter 5: The Design Process

This chapter provides an overview of the user-centred design process that incorporates users throughout the different stages of the design. The four stages of the design process include the generation of design recommendations, designing, modelling and evaluating. These stages are briefly introduced in this chapter before each one is focused on in more detail across the remaining chapters of this book.

1.5.6 Chapter 6: Design Requirements

This chapter provides detail on the initial starting point for developing and designing new technologies; generating the design requirements. It focuses on two approaches to generating requirements. The first is in relation to advancing current systems, which includes developing an existing system and/or altering or improving it in some way. A case study example is provided of an existing flight management system which was reviewed to determine requirements for improved design. The second approach is the design of new technologies that have not yet been developed. Here the requirements must outline what the new technology will be utilised for and how it will be integrated with the other technologies within the current system. Another case study example of a new pilot decision-making aid is presented, with design requirements gleaned from interviews with commercial airline pilots.

1.5.7 Chapter 7: Design Generation

This chapter presents an approach to developing the designs for new technologies on the flightdeck with user involvement. It presents the Design with Intent method which enables end-users to take part in design workshops to discuss and draw out their design ideas, while also focusing on key design principles. A case study example is presented which relates to the pilot decision-making aid that was initially presented in Chapter 6, and for which design requirements were generated. The second part to this chapter then compares the design outputs from the end-user workshops to requirements generated from a more traditional Human Factors approach that focus on minimising error in the design of new technologies. Comparisons between these two approaches highlight the benefits of including end-users as well as Human Factors practitioners within the design process.

1.5.8   CHAPTER 8: DESIGN MODELLING

Once the designs have been outlined, the next step in the design process is to model how they will be used. This chapter provides two approaches to modelling new designs. The first modelling approach is engineering integration modelling which considers the functions and tasks that a system will perform and how new technologies will integrate within these systems. Operator Event Sequence Diagrams (OESDs) are suggested as a beneficial modelling technique to present how the user and the technology will perform within the wider functioning of the system. The second approach is end-user modelling, which focuses more on the cognitive processing of the human user and how their decision-making and thought processes may be impacted by the newly designed technologies. Here different decision-making models are presented that highlight how new decision tools on the flight deck may alter the pilot's cognitive processing and the different ways that new designs need to support the end-user.

1.5.9   CHAPTER 9: DESIGN EVALUATION

This chapter presents the final stage in the design process, which is evaluating the designs. Three different areas of evaluation are discussed in this chapter which consider the different level of fidelity and the context of use which are important to the evaluation stage. Low-fidelity heuristic evaluation methods are discussed which require wire frame mock ups only and can be used to achieve early evaluation. Higher fidelity simulation testing is then discussed with an example from a flight simulator study which includes a sample of qualified airline pilots. Finally, the consideration for context of use is discussed with respect to evaluating future cockpit interfaces under turbulent conditions. Here a study on a motion platform with different types of touchscreen interfaces is presented to demonstrate different Human Factors considerations when designing future flight deck technologies.

1.5.10   CHAPTER 10: CONCLUSION

This concluding chapter provides a summary of the work presented across the chapters of this book. It presents the main contributions in relation to the user-centred design process with respect to future cockpit design. It also highlights the importance of end-user involvement and Human Factors principles to design in the aviation domain.

2 Human Factors Approach to Research and User-centred Design

2.1 INTRODUCTION

This chapter will introduce you to different elements of experimental or research design that should be considered when conducting a study with human participants. There are a number of reasons why you might want to undertake a Human Factors study, including; to test or find out if a hypothesis (research question) can be supported (or not), to explore the relationship between different variables, to develop or validate models, and to evaluate features of design from single interfaces or products through to whole systems. Whether you are conducting a naturalistic observational study in a 'field' setting or a very controlled high-fidelity simulator experiment, the same factors will need to be considered and their choices justified (Section 2.2). Human Factors research primarily involves human participants and seeks to assess at least one component of cognitive or physical/biomechanical behaviour or performance in complex systems. Section 2.3 will explain what we mean when we talk about user-centred design, why it is important and how you can ensure you are implementing it in your design processes. Finally, Section 2.4 will provide a brief introduction to research ethics, why they are important and how you can ensure you are conducting ethical research.

2.2 RESEARCH DESIGN CONSIDERATIONS

This section is intended to highlight the most common considerations that need to be made when planning a study. Every study will follow a different experimental design protocol, so it is not possible to define or prescribe the specific parameters of individual studies, rather this section will present a number of factors that should be considered to ensure your study is as accurate as possible. Traditionally, experimental design primarily referred to conducting objective and controlled research in laboratory settings. In this context, precision is maximised and statistical analyses would determine where significant results were obtained in the data. This approach might be relevant for a simulator study with highly controllable parameters and a clear relationship between cause and effect (**Chapter 9** at the end of this book will cover this type of research), but Human Factors research in safety-critical domains like aviation can often be more applied in nature. This might result in less experimental control, uncertain hypotheses

DOI: 10.1201/9781003384465-2

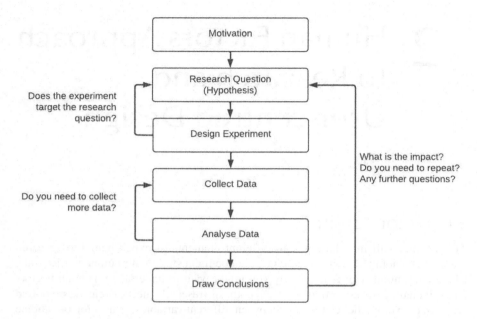

FIGURE 2.1   Research design process for Human Factors research.

(expected outcomes) and produce qualitative data that do not lend themself to statistical analysis. As such, we are using the term 'research design' to refer to the considerations that need to be made when planning a study. Figure 2.1 depicts the research design process that generally any study involving human participants will follow. A study begins with a motivation, that is, the reason something is being investigated. This might be a personal motivation of interest or a client directive. This will lead to the generation of a research question (or multiple questions), which is needed to guide the research design considerations (i.e., justify choices and trade-offs in terms of methods, participants, setting, etc.). Once a study has been designed it is highly advisable to conduct a pilot (preliminary) study to ensure that the research question is being addressed. If not, it might be that the study has not been designed appropriately or in fact the wrong research question is being asked. Research design is therefore an interactive process with feedback loops to refine the design throughout the process. After a pilot study and any re-design, data can be collected and then analysed, which may lead to additional data collection and will eventually enable conclusions to be drawn. This invariably leads to the generation of additional research questions which might warrant future investigation. The remainder of Section 2.2 will explain the components of research design in more detail.

2.2.1   Generating a Research Question (or Hypothesis)

Research questions or hypotheses are the initial building block of the scientific method and the starting point of a Human Factors study. They are similar, but not the same. Typically, a research question is a concise and focused statement defining what the study aims to answer and as such, provides a clear path for the research process (see Figure 2.1). They can be descriptive (a study that is designed to describe what is going on), relational (a study that is designed to look at relationships between factors) or causal (a study design to test cause and effect between different variables). A causal study usually involves generating hypotheses, which is a formal, testable, statement, usually informed by previous research findings, designed to predict the relationship between two or more variables (see Section 2.2.2). The purpose of a study is to support or refute a hypothesis, rather than to *prove* it. The different types of hypotheses are listed below:

- *Null hypothesis*: states there is no relationship between the two variables being studied (e.g., there will be no difference in pilots' situation awareness ratings using a head up display and not using one).
- *Experimental (or alternative) hypothesis*: states that there is a relationship between two variables being studied (e.g., there will be a difference in situation awareness ratings between using a head up display and not using one).
- *Directional hypothesis:* is a type of experimental hypothesis that will predict the nature of the effect (e.g., situation awareness will be *improved* with a head up display, this is a 'one-tailed' hypothesis as it states the direction of the effect) and a *non-directional hypothesis* will not (e.g., there will be a difference in situation awareness, and is therefore a 'two-tailed' hypothesis).

2.2.2 Variables

In order to write a hypothesis or research question, the variables under investigation have to be defined. They include anything that can change or be changed.

- *Independent variable* (IV): is the variable that is being manipulated or changed within the study. For example, this could be whether a flight scenario is performed with or without a head up display or type of weather condition (good or poor visibility). A study may have more than one IV, but usually no more than two or three otherwise the cause and effect relationships become too hard to differentiate.
- *Dependent variables* (DV): are the variables that are measured (it is assumed that the IV affects the DV). It is common to have multiple DVs, these could include elements of human performance (e.g., types of errors, task time and workload scores).
- *Confounding or extraneous variables:* Are factors that might impact the results of a study if they are not controlled for. They can be situational (i.e., the environment the study is conducted in), person-based (i.e., people have good and bad days and are different from each other), experimenter-based (i.e., the presence of a researcher might influence the performance of participants) and demand characteristics (i.e., people want to please and do well on studies which may impact their performance). Confounding variables can be controlled for by the type of design used (see Section 2.2.3) and following a standardised procedure (see Section 2.6).

2.2.3 TYPE OF STUDY DESIGN

The study design is a very important consideration and will impact other elements of the research including participant recruitment (see Section 2.2.4) and procedure because it defines how the research is structured. There are two main types of research design, each with associated strengths and limitations:

- *Independent (or between) measures:* Different participants are used in each condition of the IV, so each condition has different groups of people participating (e.g., two groups of participants test two different head up displays). The advantage of this is that it reduces practice effects (i.e., if a person is involved in repeated trials, then they can become bored or savvy to the requirements). However, more people are needed, and between-group differences may affect the results (and become a confounding variable, see Section 2.2.2). A way to control for this is to use a *matched pairs* design, whereby participants are different for each condition but participant variables are controlled by matching to a key factor, for example, flying hours or qualification.
- *Repeated (or within) measures:* uses the same participants across multiple conditions (e.g., one participant will test three different head up displays). The advantages are that fewer people are needed and the between-group differences are limited. However, practice effects may become more apparent (i.e., if the head up displays were presented in the same order for all participants, their performance might be the result of which one they tested first or last, rather than the design of the display). Therefore, *counterbalancing* is used to reduce this. This involves changing the order that different conditions are presented to each participant (i.e., participant A tests head up display 1, 2, 3, whereas participant B tests head up display 2, 3, 1). Practice effects may still occur for each participant, but they are equal in both groups and so cancel each other out.

2.2.4 PARTICIPANTS

Participants should be selected in accordance with the research question(s). For example, there is no point in using a large pool of student participants for a study of novel cockpit concepts tested in a simulator if the student participants cannot fly the simulator and have no domain knowledge. It is generally accepted that studies involving subject matter experts will involve fewer participants as access to them can be limited and resource intensive. What is lost from fewer participants in terms of statistical power is usually compensated by their domain expertise and the insights that can be gained. Virzi (1992) found that 80% of usability problems (including the most severe problems) were detected in the first four–five subjects and advocated that smaller and less costly studies produced comparable results to similar but larger studies. The important factor with participant selection is gaining a *representative sample*. For example, for aviation research, the demographics of the participants should mirror the demographics of current pilot populations and the expected demographics of future populations (in terms of gender balance, age profiles, nationalities, etc.). It is usually impossible to study every single person in a target population so a sample or sub-group of the population is sought, that is likely to be representative of the target population of interest. Different sampling (recruitment) approaches include:

- *Opportunity sampling:* when members from the target population are chosen simply because they are willing to be participants in the study (volunteer). This is perhaps the quickest way of going about selecting participants but is likely to be unrepresentative.
- *Random sampling:* When the target population has been identified and participants who will become part of your study are randomly selected (e.g., a random selection of all trainee pilots at a certain airline). This can be done by allocating numbers and using a random selection computer program. Each member has an equal chance of being selected and there are no biases in the selection process (though it might be necessary to acknowledge biases that exist in the target population, e.g., pre-existing gender biases). It can, however, take a lot of time to set up and requires buy-in from employers and has ethical implications around pressure to participate (see Section 2.4).
- *Systematic sampling:* Every *n*-th member of the population is selected as a participant for example, every 20th. This is a simple procedure but may not necessarily end up as a representative sample.
- *Stratified sampling:* The researcher selects an equal percentage of participants from each subgroup of the target population. This is the most representative of the sampling techniques but can take a long time to set up.

Before agreeing to be involved in a research study, participants should be given some details about the requirements, usually in the form of a 'participant information sheet' (e.g., why the research is being undertaken, time involved, location, any risk of harm such as motion sickness from a simulator, any incentives like expenses payments or refreshments). More detail on interacting with participants is included in Section 2.4: Research ethics. During the data collection phase of research, it is good practice to collect relevant *demographic* information about the participants (e.g., age, gender, job role, years of experience, etc.). This can be used in the analysis to report the sample characteristics and acts a basis for comparing results against.

2.2.5 Measures and Equipment

A *method* is a particular procedure for investigating something (e.g., subjective self-report questionnaires) and *measures* are the tools or instruments for assessing the metrics (or DVs) of performance (e.g., the NASA-TLX measure to assess mental workload). Numerous factors will influence what measures are chosen in your study, which might include:

- Stage of design life-cycle (some measures lend themselves to more mature prototypes than others)
- Available time and resources (cost–benefit considerations usually need to be made)

- Accessibility of the measure (is it available to use)
- Skills and expertise of researcher(s)
- Access to end-user populations (see Section 2.3)
- Type of data required (see Section 2.2.7)

Salmon et al (2009) suggested that *reliability* and *validity* should be a critical consideration when selecting methods. Methods can be reliable (provide consistent measurement) without being valid (provide an accurate reflection of a phenomenon), but they cannot be valid without being reliable (Stanton & Young, 1998, 1999). Key methods and measures relevant to the user-centred design process in aviation are detailed throughout the remaining chapters of this book, including user interviews (**Chapter 6**), design workshops (**Chapter 7**) decision modelling (**Chapter 8**) and simulator studies (**Chapter 9**). Measures are completed when a participant has interaction in an experimental (or control) condition. It is important to measure the same things after each condition and across different participants. *Equipment* refers to the physical things that are required to conduct a research study. This ranges from simple paper/pen/clipboard set-up for observational studies, to a complex high-fidelity motion simulator. It is useful to provide a technical specification of what equipment was used (including any software) so that the study conditions can be replicated, if necessary.

2.2.6 PROCEDURE

Documenting the procedure (step-by-step description of the process) is important so that a study can be understood by others and potentially replicated. It also ensures *standardisation* across the participants and contributes to reducing the impact of confounding variables. If a team of researchers are conducting a study, a common procedure should be agreed upon, documented, and followed, throughout the data collection phase. The procedure should detail:

- Instructions given to participants
- Describe what they will do in each condition and in what order (including any familiarisation time with equipment)
- How the data will be collected
- Debrief process (if applicable)

After writing the procedure, it is useful to conduct a *pilot study* which is a mini version of the full-scale study. This is conducted to evaluate factors such as feasibility, time, cost, unexpected outcomes and participant selection.

2.2.7 QUANTITATIVE AND QUALITATIVE RESEARCH

There are a wide variety of *types of studies* that can be conducted, which sit along a spectrum of control vs. ecological validity. Deciding on a type of study will depend on a number of factors similar to the considerations given for method selection (see Section 2.2.5). Different approaches have different strengths and limitations associated with them. A simulator study might offer a greater level of control than a naturalistic observational study in the 'field'. It will also be easy to replicate and standardise, and allows for cause-and-effect relationships to be established. However, it can be an artificial environment which limits the generalisability of the findings beyond the lab. The formal setting might also make it vulnerable to biases of experimenter effects and demand characteristics. At the other end of the spectrum, a field study occurs in a natural setting and is therefore likely to reflect actual behaviour, which increases its generalisability. The flexible environment might generate unexpected insights, but it does result in less control and is vulnerable to confounding variables. A field study can also be resource intensive, and observations might still result in biases and unintended changes in behaviour.

As well as a range of study types, a range of data collection options exist, which at the simplest categorisation can be defined as *quantitative and qualitative data*, each with associated strengths and limitations.

In general, quantitative research:

- Is concerned about the impact of one factor on another (i.e., IV on DV)
- Many of these comparisons can be measured quantitatively (using numbers)
- Numeric data are provided, which is useful for making predictions and replicating findings
- *Strengths*: software programs are available to help with the analysis, results are largely independent of the research so can be considered more *objective*
- *Limitations*: limited opportunities to capture emergent effects, experimental control may limit generalisability, may require an extensive number of participants to generate an effect and it can be too reductionist for understanding the complexity of interactions in complex sociotechnical systems

In general, qualitative research:

- Is concerned about gaining deep insights and understanding
- Is a broad methodological approach encompassing many research methods
- Results in unstructured textual data from methods like interviews, observations and communication transcripts
- *Strengths*: acknowledges that not everything can be reduced to numbers and provides an understanding of why and how, not just descriptive information. Provides a depth and richness of data that can highlight unexpected insights
- *Limitations*: Resource intensive and can rapidly generate a large, cumbersome data set. Open to criticism of *subjectivity* as there are various ways to analyse data, open to researcher interpretation

2.3 USER-CENTRED DESIGN

This section introduces the user-centred design (UCD) process as a founding principle when applying Human Factors methods for design, modelling and evaluation.

UCD is an iterative process, in which the end-users and their needs are placed centrally at each stage of the design process. A variety of research techniques will ensure the needs of the users are considered and the case studies throughout this book will highlight examples and best practice for UCD. 'Good' design from the start, with early involvement of Human Factors experts, is key to a successful project. The end result should be highly usable and accessible 'products' (i.e., interfaces, devices, systems).

Norman and Draper (1986) stated that 'to understand successful design, requires understanding of the technology, the person, and their mutual interaction...' (p. 1). User-centred design places the user at the centre of the design process and enforces that the equipment is designed to meet the skills, knowledge and abilities of the target user of the device (Harris, 2007). This is especially key within the early stages of the design process to account for the end-user throughout, rather than being considered when it may be too late to make adequate design changes.

Regardless of domain, the inclusion of potential end-users within the design process is essential when proposed future designs include changing the role and tasks of operators (Kaber et al, 2002). Automating specific functionality or changing the information that is presented requires users to adjust their interactions with a system or work environment (Kaber et al, 2002; Parasuraman & Wickens, 2008). Furthermore, proposed novel designs do not always consider how such systems may be adopted by end-users. Methods are therefore needed that can account for how new designs will be integrated alongside the other technologies and tasks that are being conducted within the work environment of interest.

2.3.1 THE PROCESS OF USER-CENTRED DESIGN

The user-centred design (UCD) process does not specify exact methods at each stage of the design lifecycle, rather it can be considered as an underpinning philosophy for maintaining focus on who will use the end-product. This iterative process involves users in the earliest stages of requirement generation, through to evaluations of final products. UCD consists of four general phases (www.usability.gov/what-and-why/user-centered-design.html):

1. Specify context of use (who will use the product, what will it be used for and when)
2. Requirements generation (what are the 'key performance indicators' for the product to be considered successful)
3. Create design solutions (this can be a multi-stage process, starting with paper-based prototypes and ending with final design)
4. Evaluate design solutions (through usability testing with end-users)

The UCD process is best conducted by multidisciplinary teams with complementing skills and expertise. A driving principle of UCD is the need to understand end-users, the tasks they will undertake and the environments in which they will work in. This is the approach we apply throughout the following chapters of this book.

2.3.2 DESIGN STANDARDS

The International Standard 9241 is the basis for many UCD methods (International Organization for Standardization, ISO, 2019). The standard, *Ergonomics of human-system interaction – Part 220: Human-centred design for interactive systems*, provides requirements and recommendations for human-centred design principles and activities throughout the life cycle of computer-based interactive systems. It is intended to be used by those managing design processes, and is concerned with ways in which both hardware and software components of interactive systems can enhance human–system interaction. ISO 9241–220 provides a comprehensive description of the processes that support the activities that are required as part of human-centred design, structured around human-centred processes (HCP) at various levels of an organisation:

- *HCP.1* addresses what organisations need to do to enable human-centred design on a corporate level,
- *HCP.2* describes the required infrastructure and management of human-centred design across projects and systems,
- *HCP.3* details the project-specific aspects of human-centred design during development or change of a system,
- *HCP.4* covers the specific processes during introduction and operation of a system.

Together, the implementation of these four sets of processes ensures that the systems produced, acquired and operated by an organisation have appropriate levels of usability, accessibility, user experience and mitigation of risks that could arise from use.

2.4 RESEARCH ETHICS

Research ethics relate to the integrity, transparency and quality in the design and conduct of research. Activities which involve human participants should give some consideration to ensure that basic principles of ethical practice are followed. These principles and ways in which they can be achieved are highlighted below:

- *Voluntary participation and the right to withdraw*: people should not feel pressured or coerced into participating in a study. There should be no conse-quence to them declining participation (e.g., to working relationships or oppor-tunities), nor should they feel obliged to continue in a study once they have started. Participants should also have the right to withdraw their data after a study has been completed. It is good practice to draft a 'participant information sheet' (PIS) at the start of a study to clearly highlight their right to withdraw at any time.
- *Informed consent:* where possible, participants should be informed of the objectives of the study and manipulations they will be exposed to. If this is

not possible (because it will impact performance and results) then participants should be *debriefed* at the end of the study as to the nature of the research. It is good practice for participants to sign a consent form, which states that they are agreeing to take part in the study and any other special features of the study (e.g., video recording, or what will happen to the data).

- *Avoidance of harm*: participants must leave a study in the same state of mind as when they entered, and risk of harm (physical or psychological) should be no greater than one could expect to experience in their normal course of life. A comprehensive PIS should detail any potential risk of harm and mitigation strategies in place, this will ensure a participant has been able to make fully informed consent before deciding to take part in the study.

- *Anonymity and confidentiality*: Research can visit sensitive topics, measure performance and expose participants to expressing views outside of day-to-day life. Researchers have a moral duty not to divulge information relating to a specific individual, with data and reports being anonymised and un-identifiable. Measures taken to ensure confidentiality and anonymity should be explained to participants in the PIS before consent is given.

2.5 CONCLUSIONS

This chapter provided an overview of things that need to be considered in research design. Section 2.2 detailed the considerations that should be given to different elements of the research process, including: generating research questions, variables, study design, participants, measures, and equipment and procedure. Options discussed here should be decided upon and justified in the context of the research question(s), as the case studies throughout this book will demonstrate, there is not a 'one-size-fits-all' approach when it comes to making research design considerations. Section 2.3 briefly introduced the user-centred design process which is central when applying Human Factors methods for design, modelling and/or evaluation. Keeping the end-user at the very heart of this process will ensure a successful end-product, as measured by factors including: meeting user requirements and expectations, minimising error potential (low frequency and severity), maximising efficiency, optimising workload, resulting in user satisfaction and usability. Finally, Section 2.4 highlighted the key principles of ethical research and how these can be achieved in practice. The next chapter will explore which Human Factors methods are most suited to the design process. This will include how to identify which methods may be appropriate for your research questions and how to apply them.

3 Human Factors Measures and Methods[1]

3.1 INTRODUCTION

This chapter will detail how the application of different measures and methodologies can facilitate the integration of Human Factors into the developmental lifecycle of new avionic technologies. Selecting the 'best' or most suitable method can be a challenge. From a practical point of view, method selection is typically based on four specific criteria: time, cost, expertise and equipment. The intrinsic detail relating to each of these criteria is, however, highly individualistic depending on the company and project. The constraints and the types of methods available will also vary depending on the stage of the design process. This chapter will provide an overview of the different aspects of human performance that can be assessed with Human Factors methods and that are relevant to the aviation domain. It will then go on to provide a summary of some of the key design and evaluation methodologies that have been applied over the course of the aviation projects presented within this book. These methods will also be evaluated to present their benefits alongside their challenges and limitations.

3.2 SUBJECTIVE MEASURES

As the world recovers from the impacts of the COVID-19 pandemic, air traffic is returning to, and in some cases, surpassing previous highs. Increasing levels of air traffic is in turn leading to heightened levels of complexity and inter-connectivity within the aviation domain, including both in-flight and on-ground activities. This increasing complexity therefore demands a focus on the optimisation of human performance in the piloting and management of aircraft. The central role of the human in maintaining operational safety is critical and should therefore be closely assessed and managed to ensure pilots are performing effectively within the air transport system. Any changes to the role of human operators within the aviation system need to be meticulously measured, assessed and monitored to ensure that human performance remains within acceptable safety bounds. Objective measures can directly measure task performance, but they are unable to assess an individuals' subjective interpretations of their own experience, which is an integral component of human behaviour and can offer insights into how behaviours can be improved. Subjective, self-reported

DOI: 10.1201/9781003384465-3

TABLE 3.1
Research Approaches and the Example Data They Generate, Adapted from Parnell et al. (2018)

	Quantitative	Qualitative
Objective	Measuring and quantifying behaviour e.g., eye tracking measurement, reaction time	Observing measures of behaviour, e.g., tick-box response to a survey
Subjective	Measuring views of behaviour, e.g., Likert scale in a questionnaire	Observing views of behaviour, e.g., open-ended interviews

measures are therefore vital in substantiating understanding alongside objective performance data.

There are a variety of methods that can be used to study human behaviour. Broadly these can be categorised between two key dichotomies: quantitative/qualitative and objective/subjective dichotomies. The quantitative/qualitative dichotomy has been introduced in Chapter 2 (Section 2.2.7). Within the second, objective/subjective, dichotomy, subjective data relate to the individual's personal judgements and opinions, whereas objective data involve impartial measurement of established metrics, such as physiological output, for example heart rate. These objective/subjective research foundations feature in both qualitative and quantitative research. The implications of subjective, objective, qualitative or quantitative methodologies that are used to study behaviour are important to understand as they inform the type of data that can be collected and the knowledge that can be obtained in experimental settings. Examples of the type of data are illustrated in Table 3.1.

The role of the human within the aviation domain, as in many other domains, is evolving with the development of automation. The impact that this shift has on individuals' views of their own behaviour and performance during interactions with modern automated systems are therefore of great importance. Through the consideration of an individuals' views of their own behaviour, their needs and desires can be reviewed, and this understanding can be used to support user-centred design for newer systems. The additional consideration and inclusion of human performance metrics into the design of the flight deck is essential to enhance performance and maintain safety. It is paramount that the introduction of new technologies does not introduce new unforeseen problems. Thus, the use of performance measures to assess the impact of new technologies integrated into the flight deck will be vital in ensuring that they do not adversely affect the performance of the individual or the system as a whole.

3.3 SUBJECTIVE HUMAN PERFORMANCE MEASURES

There are numerous aspects of human performance that can be assessed and measured. Some are better measured using objective methods that can quantify aspects of behaviour such as reaction time or error rates. Others, such as acceptance, can be better understood using subjective methods. Whilst objective measures can be directly assessed, understanding the range of potential subjective data which can be included within research is essential to develop valuable and informative research studies. The diverse range of performance measures that can be assessed with subjective methods, and that are relevant to and frequently used within the domain of aviation, were therefore sought through consideration of published academic literature. Sixteen key areas emerged from the literature search. The measures are described below.

3.3.1 USABILITY

The definition of usability is deemed to be intrinsically related to the context of use in which an object is being assessed (Harvey et al, 2011; Ophanides & Nam, 2017). There have, however, been attempts to define usability. For example, the International Organisation for Standardisation (ISO) 9241 Ergonomic Requirements for Office Work with Visual Display Terminals Part 11, Guidance on Usability (ISO, 1998) states that usability is '[the] extent to which a product can be used by specified users to achieve specified goals with effectiveness, efficiency and satisfaction in a specified context of use' (1998, p. 2). Subjective measures of usability have been used to inform the design of interfaces to enable them to be functional and readily adopted by the user. They are also useful once interfaces are operational, to evaluate how they may be impacting on performance. While objective measures of error rates can identify the accuracy of performance when interacting with new technologies, the subjective element of usability can assess the individuals' level of understanding and the practicality of the technologies within the environment they are intended. Feedback from these measures can inform specific design alterations and suggest improvements.

3.3.2 USER ACCEPTANCE

This relates to the willingness of the user to employ the system under assessment for its intended purpose. Thus, methods seek to explore how systems will be adopted by the user and their subjective perceptions of its use. User acceptance is of critical importance to determine if systems could be rejected by the user population. Understanding how systems and tools may be designed and adapted to be more widely accepted is critical in order for systems and tools to be used for their intended purposes. As novel technologies and applications are implemented within the flight deck, it is important to understand how these will be accepted by the user population. As individuals may find it difficult to elucidate why they prefer one system over another or indeed may intrinsically prefer one system over another, these measures are predominately subjective in nature.

3.3.3 USER CONFIDENCE

The confidence that a user has in the system relates to its perceived reliability, consistency with expectations and its reactions to the users' interactions with it (Corno et al, 2015). This relates to user acceptance (detailed above) as the more confidence that a user has in the system, the more likely they are to accept it. User confidence is however more focused on the processes integrated into the system and their function-ality, as well as the users' own perceptions on how they come to adopt and rely on the system. Within the flight deck it is important that the pilot has confidence in the technologies that they have available to them and that they can rely on the information that the system is presenting. This is only something that can be effectively applied subjectively by obtaining the user's point of view.

3.3.4 TRUST

Trust relates to the two aspects of human performance previously considered: user acceptance and user confidence. The assessment of trust in systems from a Human Factors perspective has become increasingly important with the developments in automation (Lee & See, 2004). Users are now thought to respond to automated technology through interactions that are similar to social interactions with another human, with trust being recognised as a large component within social interactions (Reeves & Nass, 1996; Lee & See, 2004). For automation to be effective, strong and trusting relationships must be formed between the user and the technological system. Ineffective human–automation relationships can lead to misuse or disuse of auto-mation (Parasuraman & Riley, 1997). The misuse of automation within aviation has resulted in fatal incidents, for example due to over-reliance on auto-pilot (Sparaco, 1955). Subjective ratings of trust in systems can inform us of how individuals may come to be reliant on functionalities and automated capabilities and therefore inform design requirements for future systems.

3.3.5 MOTIVATION

Technology is now central to the attainment and pursuit of goals. As motivation is important to goal attainment, Szalma (2014) states that motivation is a key driving force in human–technology interactions. Hancock (2013) argues that it is the design of the task that leads to a lack of motivation, rather than the actual task itself. Thus, assessment of the users' motivation to use and perform tasks through interaction with the system is an important aspect of human performance to measure. This is a facet of behaviour that can be best measured subjectively, to be understood from the users' viewpoint.

3.3.6 SITUATION AWARENESS (SA)

This relates to the level of awareness that an individual has for the environment that they are located within and the situation that they are engaged in. To be successful in performing a task it is vital that the individual has a good level of understanding of their current environment. Situation awareness can be measured through physio-logical measures and objective performance measures. Yet, subjective measures are useful as they are predominantly inexpensive and unobtrusive in nature and enable assessments in both simulated and real-world environments (Jones, 2000). Subjective methods of assessing SA have been an important development in aviation research methodologies (e.g., Waag & Houck, 1994; Taylor, 1990; Vidulich & Hughes, 1991). Despite the effectiveness and the insight offered by measures of situation awareness, they have been criticised for frequently occurring post event, and therefore rather than being a real-time assessment, they are often a recollection of events and their context.

3.3.7 Task Performance

The performance of specific tasks can be readily assessed in an objective manner, through error counting or reaction time. Yet, subjective measures of task performance can be used to assess anticipated performance when interacting with interfaces with the aim to improve performance through making improvements to design. This has obvious practicalities in the aviation domain where errors or delayed reaction times can have major safety implications. Thus, advanced anticipation of task performance is important to obtain to inform future design requirements.

3.3.8 Human Error Identification

Human error is a concept that is heavily focused on within wide-reaching areas of Human Factors research and safety-critical operations. Human error is thought to be a key contributor to incidents in the civil aviation domain (Harris & Li, 2011). Other domains such as rail (Lawton & Ward, 2005), medicine (Helmreich, 2000) and road (Medina et al., 2004) have each claimed significant proportions of incidents arise due to human error. Reason (1990) defined human error as "all occasions in which a planned sequence of mental or physical activities fails to achieve its intended outcome and when these failures cannot be attributed to the intervention of some chance agency" (p. 9). The safety-critical nature of human error makes it essential that possible errors are identified before they occur in order to provide solutions to prevent them from becoming a reality. The use of subjective measures to determine the individual's viewpoint on possible incidents and situations which may lead to error is useful, especially when experienced end-users of the system under assessment are utilised. It is, however, important to note that human error is now not a preferred term within the Human Factors discipline (Read et al, 2021). It is now acknowledged that the term can be misleading and uninformative when other factors outside of the individual's behaviour are not considered.

3.3.9 Workload

The mental workload demanded by a task refers to the level of attentional resources that are required for its successful completion. Humans have a finite level of attentional resources that they must adequately dispense among all tasks that they are conducting concurrently (Wickens, 2002). As technology becomes increasingly utilised within systems, the level of complexity that is found within technological systems can lead to increasing levels of workload. Alternatively, increased levels of automation can lead to reduced levels of workload, to the point that users can become disengaged from the activity, known as underload (Young & Stanton, 2002). It is therefore important to measure the level of workload that a task demands and consider how this workload is perceived by the individual who is managing the system. This can be done objectively with physiological methods, which is the most accurate method for real-time workload. Subjective workload questionnaires can also be used to assess subjective workload levels at certain times.

3.3.10 FATIGUE

This refers to the decline in physical and/or mental capabilities of an individual due to excessive periods of effort, performance and/or a lack of sleep. Fatigue is thought to involve weariness that extends beyond normal tiredness. It can have detrimental impacts on many aspects of performance. In aviation, pilots operating on long-haul flights or long shifts are vulnerable to fatigue. The subjective measurement of fatigue allows the individual to state how fatigued they are feeling at a specified moment, and how this may change over time. By understanding the perceived level of fatigue, it is possible to better appreciate the impact of fatigue on performance. It should be noted that despite having a physiological basis, the experience of fatigue is subjective.

3.3.11 ATTENTION

Matlin (2004) defines attention as "a concentration of mental activity that allows you to take in a limited portion of the vast stream of information available from both your sensory world and your memory" (p. 640). As noted above, pilots are often subjected to long shifts and extended periods of time where they are regularly required to pay attention to specific aspects of overall flight performance and specific elements on the flight deck. While eye tracking and objective measures of attention can be recorded within the flight deck, subjective measures allow for the individual to describe where their attention is directed, what is driving their attentional focus and how attentive they are to specific aspects of the environment. Subjective measures of attention can therefore greatly support more objective and traditional measures of attention.

3.3.12 TEAMWORK

A team is said to comprise of two or more agents that work together collaboratively to achieve shared goals within a system (Salas, 2004). As flight deck crewing currently still requires two pilots, management of the aircraft is reliant on teamwork. Furthermore, communications to air traffic control and operations teams on the ground rely on the attainment of shared goals between different operators of aviation tasks. To understand the functioning of a team, two key elements must be understood. The first critical element to be understood is the task work which is being conducted by the team, secondly, it is vital to consider interactions between team members that allow the tasks to be completed. Measures of teamwork assess team tasks, in terms of their requirements of the team and of the task itself (Stanton et al, 2017). Cognitive aspects of team task analysis can also be obtained through methods that seek to determine decision-making processes of team members in a descriptive manner (Klein, 2000).

3.3.13 COMMUNICATION

As aviation tasks are reliant on teamwork, as discussed above, the communication within teams is an additional aspect of performance that should be assessed. The overall success of a team is heavily dependent on communication. Measures of communication can subjectively assess the content, technology use and the nature of interactions between different actors within the system (Stanton et al, 2017). This can assist in determining ways of improving communication between team members for efficient teamwork and collaboration.

3.3.14 STRESS

Stress is deemed to reflect an individual's interpretation of the mismatch between the demands of a situation and the available cognitive resources or capacity that can be mobilised to deal with it. In aviation, pilots can be subject to high levels of stress when they are dealing with non-normal or emergency situations and are frequently working within time-pressured environments. How individuals perceive and react to stress is important to assess in order to facilitate strategies to manage stress levels and identify ways in which tasks can be altered to reduce the stress associated with these activities. As the level of stress experienced is based not only on the situation that an individual finds themselves in, but also on their ability to cope, stress is a subjective measure.

3.3.15 VIGILANCE

This is heavily related to attention and fatigue as it captures the ability for an individual to pay close and sustained attention to specific situations over time, which is inherent to flight management. Determining levels of vigilance is useful in determining if the attentional resources can be effectively sustained to meet the demands of the environment.

3.4 METHODS TO ELICIT DESIGN REQUIREMENTS

When faced with complex systems, such as operations on the flight deck, it is not appropriate to rely upon a single design method or approach. Indeed, Salmon et al (2019) advocate a multi-method approach, as applying several methods to the same problem can highlight deficiencies in one approach that can be countered by others. Mixed-methods approaches have been useful in the development of design requirements for future technology (Stanton et al, 2009; Salmon & Read, 2019). Table 3.2 presents various design requirement elicitation methods that map the interactions between the user and the technological system of interest.

3.4.1 INTERVIEWS

Interviews with subject matter experts are an ideal way to begin to understand the constraints and functionalities of the system under design, as well as the wider context within which it is to be used. When designing for commercial flight decks, interviews with commercial airline pilots are strongly advised. Interviews with test pilots are also beneficial, but caution should be applied when interpreting their responses as they are not representative of the final end-user. Semi-structured interviews are the best format for end-user interview. This involves having a pre-scripted set of questions that you want to ask, whilst also being open to probe into further detail where needed within the interview. Plant and Stanton (2016) present a useful interview methodology called the Schema World Action Research Method (SWARM) which will be applied throughout this book (see Chapter 6). This is particularly useful when trying to understand the pilot decision-making process and why they behave in the way that they do.

3.4.1.1   Schema World Action Research Method (SWARM)

The Schema World Action Research Method (SWARM) is an interview methodology that was specifically developed to understand aeronautical critical decision-making in relation to the Perceptual Cycle Model (PCM) (Plant & Stanton, 2016). The method provides a taxonomy of the three key features of the PCM framework: Schema, Action, World (SAW). Plant and Stanton (2016) utilised transcriptions from pilot discussions on critical aviation events to identify six Schema themes, 11 Action themes and 11 World themes relevant to the management of critical aviation events. These comprise the SAW taxonomy. Each theme has a number of interview prompts that allow interviews to be conducted with pilots to extract information for the development of a PCM. There is a total of 95 prompts, however, they are comprehensive, and not all prompts are relevant to every event, so down-selection is advised (Plant & Stanton, 2016). See Plant and Stanton (2016) for a full list of all available prompts. This method is applied with numerous examples throughout the later chapters of this book.

3.4.2   Operator Event Sequence Diagrams (OESDs)

The Operational Event Sequence Diagram (OESD) method, also sometimes referred to as Event Sequence Diagrams (ESDs; Swaminathan, & Smidts, 1999) or Operational Sequence Diagrams (OSDs; Johnson, 1993) was originally developed to map machine decision sequences (Kurke, 1961). The method is now used however to illustrate complex multi-agent tasks (Kirwan & Ainsworth, 1992; Sanders & McCormick 1982), that is to say, it visually maps the role of people and technology operating within the same system (Stanton et al, 2013). In essence, OESDs are interaction diagrams between two or more agents that enable the visualisation of task allocation as a function of time. Previous research (Sorensen et al, 2011) has used OESDs specifically within an aviation context, furthermore, Harris et al (2015) advocated the use of OESDs when exploring the potential adoption of a distributed crewed flight deck. Huddlestone et al (2017) have gone so far as to begin to map the technological requirements and initial task allocations required for distributed crewing. As its name implies, the output of an OESD graphically presents the sequence of operational events, including the tasks performed and the interaction between operators and artefacts over time (Figure 3.1). Further information and a case study example of this method can be found in Chapter 8.

3.4.3 HIERARCHICAL TASK ANALYSIS (HTA)

Hierarchical Task Analysis (HTA; Annett et al., 1971) is used to decompose a task under consideration into its simplest operations. HTA is a powerful technique because it forms the starting point for many other complex HEI methodologies (Parnell et al., 2019). HTA representations are essentially comprised of a hierarchy of goals. These goals are broken down into operations. Each operation is then broken down into plans to provide an indication of task sequence and flow (Annett et al., 1971). A HTA acts as the input to many other Human Factors assessment techniques including allocation of function, workload assessment, interface design and evaluation. In its most basic form, a HTA provides an understanding of the tasks required in order to perform and achieve certain goals. According to Harvey et al (2011), a HTA can be used to assess both the efficiency and effectiveness of systems design and should be used early on in the design process, often as a precursor to other methods (Figure 3.2). Further information and a case study example of this method can be found in Chapter 6.

3.4.4 SYSTEMATIC HUMAN ERROR REDUCTION AND PREDICTION APPROACH (SHERPA)

The Systematic Human Error Reduction and Prediction Approach (SHERPA; Embrey, 1986) works on the assumption that errors can be predicted when reviewing

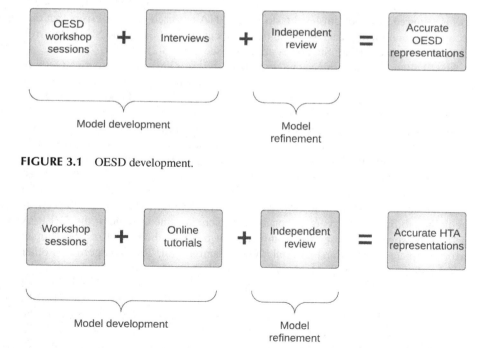

FIGURE 3.1 OESD development.

FIGURE 3.2 HTA development.

TABLE 3.2
SHERPA Error Taxonomy

Error categories	Error sub-categories
Action	Operation mistimed (A2)
	Operation too much/too little (A4)
	Right operation on wrong object (A6)
	Wrong operation on the right object (A7)
	Action omitted (A8)
Checking	Check omitted (C1)
	Right check on the wrong object (C3)
Selection	Selection omitted (S1)
	Wrong selection made (S2)
Communication	Information not communicated (I1)
	Wrong information communicated (I2)
	Information communication incomplete (I3)
Retrieval	Information not obtained (R1)
	Wrong information obtained (R2)
	Information retrieval not complete (R3)

the tasks to be performed (Baber & Stanton, 1996; Stanton & Baber, 2002; Stanton & Stevenage, 1998). In order to conduct a SHERPA, an accurate HTA of the scenario is required. The bottom-level tasks of the HTA are used, and the errors that can occur when completing each of these tasks are categorised using the SHERPA error taxonomy (Embrey, 1986; Stanton et al., 2013; see Table 3.2).

The likelihood of occurrence and severity of the error are also considered. Outputs should then be sent for independent review. Each error is considered in turn and remedial measures re-proposed in an effort to reduce the likelihood of errors from occurring. This process is presented in Figure 3.3.

Stanton et al (2002) presented the validity and reliability of SHERPA in the aviation domain. They conducted an inter-rater reliability assessment over a period of 4 weeks and found that aviation novices could apply the method with relative ease and to an acceptable level of performance when compared to airline pilots; thus, motivating its use by Human Factors researchers in the domain, alongside others who found evidence for its concurrent validity (Stanton & Stevenage, 1998) and error-predicting abilities (Stanton & Baber, 2002). There is some critique of the SHERPA method, however, by those who comment that other methods such as the Human Error Template (HET; Marshall et al, 2003; Stanton et al, 2006) outperforms other error-prediction methods due to its increased level of accuracy and ease of application (Stanton et al, 2009). The HET method is a flight deck certification tool, so it does not seek to reduce errors and provide remedial measures in the same way that SHERPA does. Harris et al (2005) argue that the SHERPA is the best HEI tool to use in the aviation domain to inform design requirements to reduce the opportunity for errors to arise and thus improve flight safety. Further information and a case study example of this method can be found in Chapter 6 (Section 1.1) and Chapter 7 (Section 4).

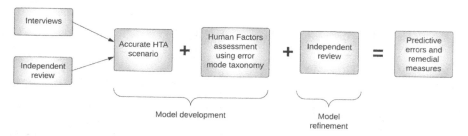

FIGURE 3.3   SHERPA development.

It should be highlighted that the choice of these different approaches to evaluating designs was driven by the same considerations as outlined previously, namely time, cost, expertise and equipment. The considerations for each of these factors are outlined in Table 3.3.

3.5   DESIGN METHODS

Human Factors in design typically refers to the ergonomic and aesthetic factors that influence the design of products and systems. Failure to observe basic ergonomic principles can have significant implications on safety, productivity and user satisfaction. Poor design contributes to work-related ill-health but more importantly has been found to be a causal factor in accidents (e.g., Ladbroke Grove; Stanton & Baber, 2008; Stanton & Walker, 2011).

There are a number of design techniques that can be used by designers early on in the design process. These include interface design techniques (e.g., Link Analysis, Layout Analysis; Stanton et al, 2013) and group design techniques (e.g., Focus Groups). Typically, the former are reliant on the availability of a pre-existing interface concept in which analysi can be performed. Group design techniques in contrast are more likely to produce insight into user needs and clarify any issues with existing system design. With this in mind, the approach used to guide and facilitate the design of interface concepts carried throughout the work conducted in the book was Design with Intent (DwI; Lockton et al., 2010).

3.5.1   DESIGN WITH INTENT (DwI)

Unlike focus groups that are based on open discussion, Design with Intent (DwI) is a form of 'directed brainstorming' involving a group of appropriate participants. It enables non-designers to design new products/interfaces quickly and efficiently, highlighting and exploring needs and desires. Unlike the Design Sprint (Banfield et al, 2015), DwI can be completed in a single session and can produce usable and innovative design plans. This is particularly advantageous given the constraints associated with recruiting commercial airline pilots as participants (Banks et al, 2019). Thus, DwI provides a means in which discussions can be structured and compared.

TABLE 3.3
Approaches used to Help Model Task–User–System Interaction

Approach	Rationale for use	Time	Expertise	Resource and equipment required	Outputs generated
Interviews	Provide in-depth insight into the pilot decision-making process and therefore generates data that can be used to inform the development of subsequent models of task–user–system interaction	Low (for data collection) High (for analysis)	Human Factors experts Line pilots	• Interview prompts (in this case, Schema World Action Research Method; Plant & Stanton, 2016) • Facilitators (HF practitioners) • Participants (commercial airline pilots) • Recording equipment • Approx. 2 hours per interview	• User feedback on system functioning and user behaviour • Decision-making processes
Operator Event Sequence Diagram (OESD)	Widely used Systems Engineering approach used to explore the relationships between individual subsystem components	Low (for data collection) Medium (for analysis) High (for verification)	Human Factors experts Engineers Test pilots Line pilots	• Well-defined scenario • Pen/paper • MS Visio • Multi-disciplinary team (ideally HF practitioners, systems architects, commercial airline pilots)	• Comparison between 'work-as-done' versus 'work-as-envisaged'

Method	Description	Training/Application time	Experts	Tools required	Outputs
Hierarchical Task Analysis (HTA)	HTA is a popular HEI method and, due to its ability to decompose complex tasks into their simplest operations, it forms the initial starting point for many other more complex HEI methodologies (Annett et al, 1971; Stanton, 2006)	Low (for data collection) Medium (for analysis) High (for verification)	Human Factors experts Line pilots	• Well-defined scenario • Pen/paper	• Comprehensive list of sub-tasks involved in achieving a certain goal that can be used as the main input to a number of other Human Factors methods (e.g., human error identification)
Systematic Human Error Reduction and Prediction Approach (SHERPA)	Can determine possible errors in the current system; classify their likelihood and criticality before determining possible remedial measures that can be developed to prevent the errors from occurring in the future. Harris et al (2005) argue that the SHERPA is the best human error identification tool to use in the aviation domain to inform design requirements to reduce the opportunity for errors to arise and thus improve flight safety	Low (for data collection) High (for analysis) Medium (for verification)	Human Factors experts Line pilots	• Complete HTA • Error taxonomy (Embrey, 1986) • Pen/paper • MS Excel	• Report on predicted errors, their likelihood of occurrence and their severity • List of remedial measures to eradicate errors identified

The DwI toolkit facilitates creative thinking (Tromp & Hekkert, 2016) through directed mind mapping. In other words, it helps users to generate ideas and structure their thinking on particular topics (Daalhuizen, 2014). The DwI toolkit is therefore marketed as a 'suggestion tool' and is comprised of 101 individual design cards, covering eight different design lenses:

- Architectural – prompts discussion on how we can use the structure of a system to influence behaviour;
- Error proofing – prompts discussion on how we can use different approaches to try and limit or mitigate the risk of errors occurring;
- Interaction – prompts discussion on some of the most common design elements in which a user's interaction with the system can affect their behaviour;
- Ludic – prompts discussion around different techniques that can be used to influence behaviour that are derived from games or otherwise 'playful' interactions;
- Perceptual – combines numerous ideas about how users perceive patterns and meanings as they interact with systems. This challenges people to think about how patterns may be used to influence behaviour;
- Cognitive – prompts discussion on how people make decisions and how such decisions are influenced by heuristics and biases;
- Machiavellian – essentially represent design patterns that are often considered unethical, but typically used to control and influence behaviour. Examples include advertisement and pricing structures; and
- Security – prompt discussion about ways in which users can control their habits or behaviours that are beneficial to the self.

These lenses provide different 'worldviews' on behaviour change and therefore challenge designers to 'think outside the box'. Each design card incorporates a provocative question that is deliberately designed to prompt detailed deliberations. The cards essentially act as a source of inspiration to trigger innovative and creative design thinking (Eckert & Stacey, 2000).

The toolkit is highly flexible in its application. Firstly, facilitators can choose between two different modes of application. Alternative inspiration mode utilises all 101 design cards, whereas the prescriptive mode of applicable enables the down-selection of cards based upon facilitator requirements. There are also a number of procedures that can be followed, including a lens-by-lens approach, selecting random pairings and identifying target behaviours. More guidance and information can be found on the DwI website (www.designwithintent.co.uk). It is important that individual facilitator teams discuss and agree on a chosen approach.

Further information and a case study example of this method can be found in Chapter 7. Table 3.4 provides a summary of this design methodology.

3.6   EVALUATION METHODS

There are a number of Human Factors methodologies that can be used during the evaluation of a product or technology. The approaches used within the current work

TABLE 3.4
Approaches used to Help Design Interface Concepts

Approach	Rationale for use	Time	Expertise	Resource and equipment required	Outputs generated
Design with Intent (DwI)	DwI is a directed brain-storming activity that provides some degree of structure to discussions. It enables non-experts to design new products/interfaces quickly and efficiently. Unlike the Design Sprint (Banfield et al., 2015), DwI can be completed in a single session and produce usable and innovative designs. This is particularly advantageous given the constraints associated with recruiting commercial airline pilots as participants	High (for data collection) Low (for analysis) Low (for verification)	Human Factors experts Line pilots	• Well-defined scenario • DwI cards • Pen/paper • Recording equipment • Approx. 1 day (for all 101 cards)	Novel interface design concepts

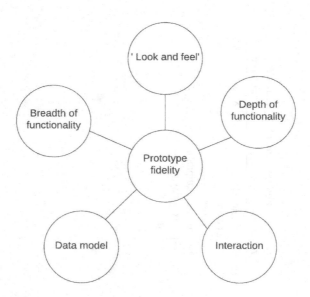

FIGURE 3.4   Five dimensions of prototype fidelity.

were Heuristic Evaluation, Desktop Simulation and Laboratory Studies (Table 3.5; for further explanation see **Chapter 9**). It should be noted that only during the latter stages of development can the subjective measure highlighted previously be considered for inclusion in testing.

Typically, in the evaluation phase of the design lifecycle, a distinction is made between low- and high-fidelity test prototypes. Low-fidelity prototypes offer zero to limited functionality and can therefore be developed at much lower costs and earlier on in the development lifecycle. In contrast, high-fidelity test prototypes provide a much more representative 'look and feel' scenario in which detailed interactions can be assessed (Rudd et al, 1996). Previous models of fidelity in usability testing have typically included five dimensions, as presented in Figure 3.4 (McCurdy et al, 2006; Snyder, 2003; Virzi et al, 1997 for more information).

In order to explore the behavioural implications of novel technology on the flight deck, prototypes can be developed at varying levels of fidelity. These can then be assessed using different methods of simulation, primarily desktop simulation and fixed based simulation. Desktop simulations are typically set up in a similar fashion to a personal computer and are associated with low simulator fidelity (Wynne et al, 2019). This may take the form of a simple static mock-up, such as those used within the Heuristic Evaluations (see Chapter 9 for more information on this type of evaluation). Alternatively, a single mount desktop can be used in conjunction with game-based controls to navigate through menus and provide a more realistic 'look and feel'. Whilst this approach provides a very limited field of view and lacks a wider contextual environment (i.e., a cockpit), it does provide the opportunity to gather early feedback on the usability of a display. Fixed based simulation in contrast incorporates realistic controls and cabin (i.e., cockpit), to make a far more realistic environment.

(continued)

TABLE 3.5

Approaches used to Evaluate Interface Design Concepts

Approach	Rationale for use	Time	Expertise	Resource and equipment required	Outputs generated
Desktop Simulation	Provides preliminary insight into human behaviour using an early prototype of a real-world system before the expenditure of further development and simulator trialling	High (for simulator development) Low (for data collection) Medium (for analysis)	Human Factors experts Engineers Line pilots	• Access to a working system prototype • Pen/paper	• Feedback provided from subject matter experts and/or end users can be used to further refine design
Laboratory (Simulator) Studies	Provide a means to explore human behaviour in a tightly controlled environment that offers an approximate imitation of a real-world system. Simulation can be used in many contexts including performance optimisation, safety engineering, testing, training, education, etc. Can be used to gather data regarding the physical and verbal aspects of a task or scenario	High (for simulator development) Medium (for data collection) High (for analysis)	Human Factors experts Technician/ engineer Participants (ideally line pilots)	• Simulator facility	• In the context of Open Flight Deck, simulation can provide engineers and/or designers with practical feedback on designing real-world systems

TABLE 3.5 (Continued)
Approaches used to Evaluate Interface Design Concepts

Approach	Rationale for use	Time	Expertise	Resource and equipment required	Outputs generated
Questionnaires (e.g., Workload, User Acceptance, System Usability Scale)	Offer a flexible way of collecting a considerable amount of data relating to a number of issues with Human Factors (e.g., usability, satisfaction, workload, etc.). They can be used in the design process to evaluate concept and prototypical designs as well as evaluating existing systems (see Stanton et al, 2013 for specific administration procedures)	Low (for data collection) Medium (for analysis)	Human Factors experts Participants (ideally line pilots)	• Pen/paper	Provide qualitative insight into the subjective perception of product usability, workload, acceptance, etc.
Interviews	Can capture data on pilot decision-making processing during non-routine incidents without interfering with exhibited behaviour during simulation	Medium (for data collection) High (for analysis)	Human Factors experts Participants (ideally line pilots)	• Interview prompts (in this case taken from the Critical Decision Method; Klein et al, 1989)	Provide qualitative insight into how decision-making may change as a result of technology implementation

Fixed base simulators do not however incorporate motion and so are not equivalent to a full realistic cockpit environment.

Simulations provide the opportunity to recreate operational environments that are either inaccessible, expensive to access or have high risk associated with their use, enabling early testing and development of future products within safe environments. For early product development and assessment of HMI, both desktop and fixed-based simulator configurations are therefore deemed appropriate. Once an interface idea is established, it is required to go through a series of refinements, using established methods to further develop the design. This journey, as well as the rationale for each step, is documented within Table 3.5.

3.7   SUMMARY AND CONCLUSION

This chapter has outlined how different measures and methods can help facilitate the integration of Human Factors into the developmental lifecycle of new avionic technologies. Whilst there is no 'best' method for testing current systems, or designing new ones, the large collection of methods available within the literature means that even selecting the most suitable method can be a challenge. This chapter has provided an overview of the different aspects of human performance, specifically considering the aviation domain, that can be assessed using subjective Human Factors methods. Following this, it has highlighted the key methodologies that have been applied over the course of the aviation projects presented within this book, those of Operator Event Sequence Diagrams (OESDs), Hierarchical Task Analysis (HTA), Systematic Human Error Reduction and Prediction Approach (SHERPA) and Design with Intent (DwI). Although this chapter has introduced these methods, future chapters will discuss their use and deployment in far greater depth. It should be noted that all methods that are available have associated advantages and disadvantages and it must be reiterated that whilst no method can ever be perfect, use of these established techniques can enhance our understanding of, and further improve our capabilities within the aviation domain.

Ultimately, the output of user-centred design (UCD) and good research practice is to develop a product that is usable for its intended purpose and for its intended user base. The following chapter will present an overview of what is meant by usability and how it can be defined within the context of aviation. This will provide an understanding of how usability is tightly linked to the context of use and a framework is presented for how usability within the flight deck can be reviewed.

NOTE

1 Craig Allison has contributed in writing this chapter.

4 Defining Usability Criteria

4.1 INTRODUCTION

The purpose of this chapter is to further address the current understanding and evaluation of usability in the context of aviation. The development of any technological device must be usable, yet the definition of usability has been remarkably difficult to define. It is highlighted within this chapter that usability cannot be assessed without considering the context of use. Therefore, in order to determine if new technologies within the cockpit are usable, a set of context-specific usability criteria is developed here that can be used to guide future usability evaluations within this domain. Following a review of the literature relating to the concept of usability in terms of factors, goals, design principles and standards, cockpit-specific criteria for usability are presented. The framework proposed has the potential to make a positive impact to the flight deck design process, which will ultimately deliver safer and more efficient interactions in flight deck operations.

4.1.1 WHAT IS USABILITY?

The most commonly cited definition of usability comes from the International Organisation for Standardisation (ISO) 9241 Ergonomic Requirements for Office Work with Visual Display Terminals (VDTs) Part 11, Guidance on Usability (ISO, 1998). This defines usability as:

> '[the] extent to which a product can be used by specified users to achieve specified goals with effectiveness, efficiency and satisfaction in a specified context of use'.
>
> 1998, p. 2

Here, there are three key concepts relating to the usability of a product/system:

1. *Effectiveness*: Relates to the accuracy and completeness in which a user can achieve specified goals;
2. *Efficiency*: Relates to the expenditure of resources such as human effort, cost and time in accurately completing tasks; and
3. *Satisfaction*: Relates to the freedom of discomfort and attitudes towards the use of a product.

DOI: 10.1201/9781003384465-4

However, the reason for the wide adoption of this definition is the inclusion of 'context of use' as it recognises that usability criteria are likely to differ between different operational environments and services. Context in this sense refers to the specific circumstances in which a product is used. A recognition of context is important because a designer's view of usability at the beginning of the design process could be vastly different from the expectations of the final user.

4.1.2 FEATURES OF USABILITY

A universally accepted definition of usability is lacking. This is because different situations will demand different attributes from a product to ensure optimal interaction can be achieved at all times. With this in mind, Heaton (1992) suggested that within the context of a particular product, an explicit definition of usability can be developed and that it is this definition that can be used to guide later evaluation. According to the academic literature, 'usability' is made up of factors, goals and principles. Indeed, there have been significant contributions to defining usability within the literature.

4.1.2.1 Usability Factors

The first formal definition of usability came from Brian Shackel (1986) who suggested that there were four factors of usability:

1. *Learnability*: A system should allow users to achieve an acceptable level of performance within a specified time;
2. *Effectiveness*: Acceptable performance should be achieved by a defined proportion of the target population, for a specific number of tasks within a range of environments;
3. *Attitude*: Acceptable performance should be achieved within acceptable human costs (i.e., fatigue, stress, discomfort, satisfaction) and;
4. *Flexibility*: The product should be adaptable and be able to deal with a range of tasks beyond those first specified.

Shackel (1986) was the first to propose that 'usability' could be quantified and proposed that numerical values could be assigned to each of the four attributes of usability. He argued that this would enable designers to specify exactly what performance should be reached in order to classify a product or system as 'usable'. These attributes were further extended by Stanton and Baber (1992) who included four additional, but 'equally important' factors into the definition of usability:

1. *Perceived usefulness*: The extent to which a product/system will be used;
2. *User criteria*: A system should exhibit an acceptable match between the functions provided by the product/system and the needs and requirements of the user;
3. *Task match*: The frequency at which the task can be performed and modified and;
4. *Task characteristics*: The knowledge, skills and motivations of the user population.

4.1.2.2 Usability Goals

Other definitions of usability have concentrated more upon how we can engineer products to be usable and therefore are more akin to the definition of usability goals. For example, Jakob Nielsen (1994) described five components of usability:

1. *Learnability*: The system should be easy to learn;
2. *Efficiency*: The system should be efficient to use and have a steady-state level of performance;
3. *Memorability*: The system should be easy to remember;
4. *Errors*: The system should have a low error rate and any errors should be easily recoverable; and
5. *Satisfaction*: The system should be pleasant to use.

Shneiderman (1992) also defined five Human Factors goals' against which usability could be measured:

1. *Speed of performance*: The time taken to complete a specified task, by a specified user;
2. *Time to learn*: Time taken to achieve a specified level of performance for a specified task, by a specified user;
3. *Rate of errors*: Proportion of erroneous transactions by specified users;
4. *Retention over time*: The level of performance on a specified task, by a specified user after a specified period of non-use; and
5. *Subjective satisfaction*: Subjective assessment of overall acceptability.

These goals provide the foundation in which a set of key performance indicators (KPIs) may be developed for a given product/system within a specified context.

4.1.2.3 Usability Design Principles

However, there are many other notable contributors to the area of usability. Usability design principles for instance represent 'design rules' rather than any specific usability factors or goals. Table 4.1 highlights some of the significant contributions made by various authors over the years in defining usability design principles. Donald Norman (1988, 2002) proposed seven 'principles of design for understandability and usability' of user interfaces that aimed to support the development of accurate conceptual models of interaction between a user and system/product. He recognised that the human–machine interface (HMI) is the only 'visible part of a system' and therefore must consolidate the designers' mental model with the expectations of the user. He argued that visibility of a system is based upon two components: mapping and feedback. Mapping refers to the relationship between the controls and the way in which a user interacts with them and any subsequent effect. Feedback is based upon the passage of information to the user that details the actions that have been performed following initial interaction (i.e., input). If mapping and feedback are capable of portraying an accurate representation of the product, the user is more likely to understand the system, which will improve usability (Norman, 2002).

TABLE 4.1
Usability Design Principles – Notable Contributions from the Academic Literature

Author	Design Principles	Definition
Donald Norman	Use knowledge in the world and in the head	Acknowledge that behaviour is a result of combining information in the head with information in the world
	Simplify the structure of tasks	Objects must be intuitive, accessible and usable
	Make things visible	Objects must be clear and unambiguous. Users must be able to determine an object's state and understand the alternatives for action
	Get the mappings right	Ensure that the underlying components of displays are mapped correctly (i.e., the structure of menus are intuitive), and that the controls and their effects are compatible
	Exploit the power of natural and artificial constraints	An object should suggest what it can do (i.e., incorporating affordance and forcing functions)
	Design for error	Well-designed products/systems should enable users to detect their own errors using feedback and recover from them
	When all else fails, standardise	Standardisation should enable transfer of learning from one product/system to another
Ben Shneiderman	Strive for consistency	Use identical terminology, consistent commands and sequences of action
	Enable frequent users to use shortcuts	Reduce the number and pace of interactions as frequency of use increases
	Offer informative feedback	For every action, there should be feedback
	Design dialogue to yield closure	Sequences of action should be organised into groups with a beginning, middle and end
	Offer simple error handling	If an error is made, the system should offer simple, comprehensible mechanisms to recover
	Permit easy reversal of actions	Provide the user with a means to undo erroneous actions via a single action, data entry or a complete group of actions
	Support internal locus of control	Allow users to initiate action rather than simply respond
	Reduce short-term memory load	Ensure displays are simple, allowing sufficient training time

TABLE 4.1   (Continued)
Usability Design Principles – Notable Contributions from the Academic Literature

Author	Design Principles	Definition
Patrick Jordan	Consistency	Generalisation of task requirements across different products and/or tasks
	Compatibility	Align method of operation with user expectations
	Consideration of user resources	Consider the demands on the user in terms of physical and mental resources
	Feedback	Provide meaningful indication about the results of user actions
	Error prevention and recovery	Minimise likelihood of user error and if errors do occur provide a means for quick and easy recovery
	User control	Maximise user control over the product and its actions
	Visual clarity	Considers scale, location and distance of displays for optimal performance
	Prioritisation of functionality and information	Considers the hierarchy of information – most important functionality should be easily accessible
	Appropriate transfer of technology	Considers the appropriate use of technology developed elsewhere
	Explicitness	Design cues that can convey functionality and method of operation

In contrast, Ben Shneiderman (1992) proposed 'eight golden rules of dialog design'. The benefit of Shneiderman's approach was the recognition of context in usability evaluation. Unlike previous contributors, Shneiderman proposed that applications for the home are likely to be characterised by different usability attributes in comparison to commercial applications. User satisfaction would be an important usability attribute for the former, whilst this would be less of a concern in a commercial environment because system use may be mandatory. These differences again become important when we consider KPIs for different products/systems.

Finally, Patrick Jordan (1998) proposed 'ten principles of usable design' describing different elements in design that can affect usability. Jordan suggested that usability is closely linked to 'pleasure of use', with pleasurable products yielding feelings associated with security, confidence, pride, excitement and satisfaction.

Whilst usability design principles do not necessarily assist in defining the concept of 'usability', they do contribute to a design philosophy surrounding the term. There are some obvious overlaps here, with 'strive for consistency' (Shneiderman, 1992) being similar to that of 'consistency' (Jordan, 1998). Further, 'visual clarity' (Jordan, 1998) is similar to 'make things visible' (Norman, 1988, 2002). However, they have been included in this review for completeness.

4.2 DEVELOPMENT OF CONTEXT-SPECIFIC USABILITY CRITERIA

4.2.1 UNDERSTANDING CONTEXT

Avsar et al. (2016) state that the flight deck represents a safety-critical environment where the pilots can both monitor the state of the aircraft via different flight instruments and provide control inputs to change that state. The first, controlled, sustained flight was completed in 1903 by the Wright brothers (Wright Brothers Aerospace Company, 2010). The pilot was only able to control the aircraft for 59 seconds and only covered a distance of 260 metres. At the time, there were only three flight instruments on board. As aircraft became capable of travelling higher, faster and further, there was an increased demand for more flight information. Systems in this period were predominantly analogue electromechanical or mechanical devices that all needed their own space within the cockpit. Due to the large number of (flight) instruments and controls in commercial aircraft (e.g., Boeing 314 Clipper 1938–1941), there had to be five members of crew; two pilots, a flight engineer, a navigator and a radio operator. The introduction of automation and advances in avionic systems design meant that the number of crew reduced from five to three by the 1970s (Avsar et al, 2016).

It took a further ten years before computer-based technology entered the flight deck. Modern day 'glass cockpits' present flight instruments via electrical screens that are interlinked with the Flight Management System (FMS) and autopilot functions. The FMS can be programmed to automatically follow a desired flight path and profile from take-off through to landing whilst autopilot can be used to guide an aircraft without constant 'hands on' control by the pilot (Sarter & Woods, 1995). The term 'glass cockpit' is now synonymous with multi-function displays (MFDs). MFDs are essentially a display surface that can display information from multiple sources, sometimes in several different reference frames (Mejdal et al, 2001). For example, an MFD may present groups of data (i.e., weather, terrain) separately or in combination. MFDs are becoming increasingly prevalent on the flight deck due to their design flexibility and low costs (Mejdal et al, 2001). Typically, a glass cockpit includes up to six interfaces, backup flight instruments (i.e., electromechanical instruments) and some critical systems indicators on the main instrument panel (Billings, 1997). These HMIs act as the bridge between the pilot and the aircraft subsystems.

In relation to flight deck design, the European Aviation Safety Agency (2017) published their Certification Specifications and Acceptable Means of Compliance for Large Aeroplanes (CS-25) that contains many requirements relating to system-specific design of the flight deck. CS-25.1302 specifically provides guidance and a regulatory basis in addressing design-related flight crew error. It outlines a number of design principles (similar to those of Jordan, 1998; Norman, 1988, 2002; Shneiderman, 1992) that seek to avoid and manage flight crew error. These are presented in Table 4.2.

CS-25.1302 recognises that systems design, training, licencing, qualification, operations and procedures can all impact upon safety and risk mitigation. It also recognises that design characteristics known to contribute to flight crew error have been historically accepted based on the rationale that training provision would mitigate any risk. This has, of course, been an inappropriate approach. It would also be inappropriate to expect equipment design to provide total risk mitigation. One of the

TABLE 4.2

Cockpit-specific Design Requirements for Controls and Information Displays (Adapted from CS-25.1302)

Design requirement	Definition
Controls and displays must be perceived correctly	Controls and displays must be intuitive and compatible with task completion. Information or control must be provided at a level of detail and accuracy appropriate to accomplishing the task. Controls and displays must be clear and unambiguous. Insufficient resolution or precision can impact upon the ability of the crew to perform a task adequately. Similarly, excessive resolution may make displays unreadable
Controls and displays must be comprehended in the context of the task	Controls and displays must be accessible and usable in a manner consistent with the urgency of the situation, frequency of use and duration of the task (e.g., controls used frequently must be readily accessed)
Support the crew in responding to, and performing, a task	Information should be presented to the crew relating to the impact of their control inputs on the behaviour of the aircraft. This enables them to detect and correct their own errors
Be predictable and unambiguous	The crew should know what the system is doing and why
Allow the crew to intervene when appropriate in task	The crew should be able to take action, change, alter or input into the system in a manner appropriate to the task

main benefits in defining cockpit-specific usability criteria is that they provide us with a means to assess the extent to which a product/system complies with various usability criteria. These criteria should be viewed as a means to mitigate risk and improve safety (Banks et al, 2018b).

4.2.2 COCKPIT-SPECIFIC USABILITY CRITERIA

Context-specific usability criteria have already been established in the field of driving – specifically for In-Vehicle Information Systems (IVIS; Harvey et al, 2011). Much like the IVIS criteria set list, six major factors were identified as applicable to the aviation domain.

Firstly, *dual task environment* recognises that the pilot(s) may be engaged in multiple simultaneous tasks. For example, during busy phases of flight (e.g., final approach), the crew need to complete a large number of concurrent tasks. Measuring efficiency would enable the analyst to highlight whether interaction with a system causes conflict with the primary task. Efficiency in this sense is a measure of effectiveness. If too much effort and attention are required to engage with the system, the system will be deemed inefficient because it could increase the potential risk to safety (Matthews et al, 2001). Second, *environmental conditions* recognises that internal and external conditions can have an effect on the way in which the pilot(s) will interact with a system. Displays in the instrument panel must be usable in degraded light but must also not be adversely affected by glare from the sun. They must also be usable during turbulent conditions that can cause aircraft vibration of varying degrees. Turbulence can severely impact upon the usability of controls and displays within the cockpit. For example, turbulence may lead to accidental touches, slower response times and difficulty in using controls (Federal Aviation Administration; FAA, 2011). Thus, usability can be evaluated by measuring the effectiveness and efficiency of user interaction with a product/system under varying conditions. Third, *range of users* refers to the demographic of the target population and consequentially represents a large and varied user group, even for pilots. ISO (1996) states that a diverse range of physical, intellectual and perceptual characteristics must be considered in the design and evaluation of products as it is difficult to define a specific set of user characteristics. Evaluating the effectiveness and efficiency of the interaction with a system across this range of users will be a useful marker of usability. Fourth, *training provision* recognises that pilots must undertake extensive training in relation to systems and operations in both the classroom and during simulated learning. This means that learnability is an important aspect of usability and is measured as the time taken to reach an acceptable level of performance (Harvey & Stanton, 2013). Mandatory training must be completed annually and must consider aircraft upgrades. Complex systems are likely to increase training times but systems that are intuitive generally rate highly in terms of user satisfaction and perceived usefulness (Harvey & Stanton, 2013). Landau (2002) suggests that acceptance of a system is 'extremely good' when very little learning is required. Fastrez and Haué (2008) also argue that initial effectiveness and efficiency can also be a good indicator of usability following first-time use. It is important that training is provided on individual systems to be used on the flight deck as well as training provision relating to usability of the entire flight deck. If, for example, a new decision support aid is proposed for addition onto the flight deck, pilots will need specialised and focused training upon this particular aid (e.g., how it works, what can it be used for). They will also require training on how the aid can be incorporated into their routine flight operations. *Frequency of use* recognises that the frequency of use will be affected by the exact purpose of the product and what functions it is able to perform. It is possible that not all of the functions offered by a system on the flight deck will be used regularly. For example, only one in ten landings might be completed manually (depending upon airport category). This implicates the concept of memorability, especially for functions that are used infrequently. Satisfaction is also thought to be closely linked to frequency of use. Hix and Hartson (1993) argue that short-term satisfaction will determine whether a user will repeatedly use a device. They then argue that habitual users will need to experience long-term satisfaction to ensure that they continue to use the system frequently. Finally, *uptake* recognises that uptake will be affected by user experience. Of course, there may not be much discretion as a commercial pilot given that uptake will largely be dependent upon individual carriers. Even so, 'perceived usefulness' is an important factor relating to uptake from the end-users' point of view. If, for example, an end-user is satisfied with a product, they are more likely to want to use it again (Hix & Hartson, 1993).

In order to construct cockpit-specific usability criteria, ten usability factors or goals were utilised from the academic literature (*effectiveness; efficiency; satisfaction; learnability; flexibility; memorability; perceived usefulness; task match; task characteristics;* and *user criteria*). At this stage, design principles were not considered relevant as they are not easily quantifiable. 'Error' was omitted from the high-level criteria list because, similar to Harvey and Stanton (2013), it was felt to be a contributing factor to efficiency. 'Attitude' was also omitted due to its similarity to 'Satisfaction'.

In order to translate these generic usability criteria into cockpit-specific usability criteria, they were matched to the relevant contextual factors outlined above and further iterated to identify cockpit-specific usability criteria. The process is presented in Figure 4.1. Note that 'compatibility' replaces 'task match', 'task characteristics', 'user criteria' and 'flexibility'. Importantly, each of these criteria can be measured, either objectively or subjectively, through varying means (example metrics provided

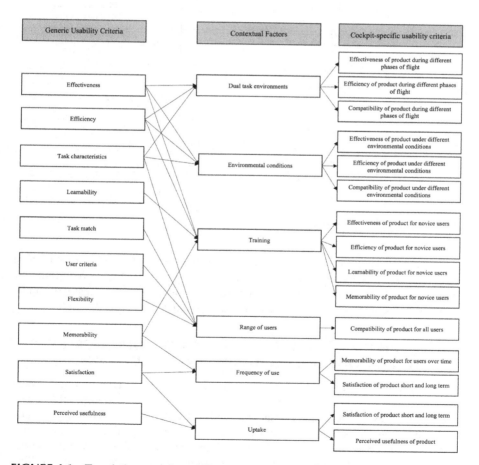

FIGURE 4.1   Translating generic usability criteria to cockpit-specific usability criteria.

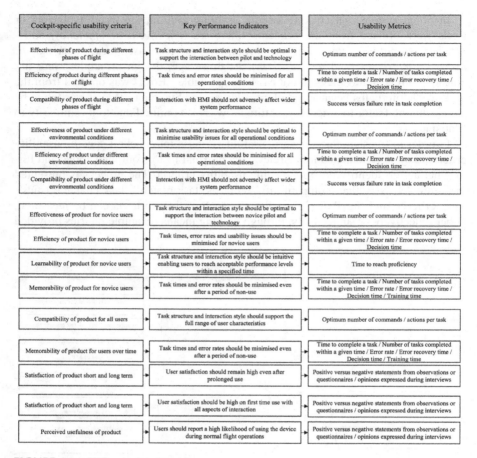

Cockpit-specific usability criteria	Key Performance Indicators	Usability Metrics
Effectiveness of product during different phases of flight	Task structure and interaction style should be optimal to support the interaction between pilot and technology	Optimum number of commands / actions per task
Efficiency of product during different phases of flight	Task times and error rates should be minimised for all operational conditions	Time to complete a task / Number of tasks completed within a given time / Error rate / Error recovery time / Decision time
Compatibility of product during different phases of flight	Interaction with HMI should not adversely affect wider system performance	Success versus failure rate in task completion
Effectiveness of product under different environmental conditions	Task structure and interaction style should be optimal to minimise usability issues for all operational conditions	Optimum number of commands / actions per task
Efficiency of product under different environmental conditions	Task times and error rates should be minimised for all operational conditions	Time to complete a task / Number of tasks completed within a given time / Error rate / Error recovery time / Decision time
Compatibility of product under different environmental conditions	Interaction with HMI should not adversely affect wider system performance	Success versus failure rate in task completion
Effectiveness of product for novice users	Task structure and interaction style should be optimal to support the interaction between novice pilot and technology	Optimum number of commands / actions per task
Efficiency of product for novice users	Task times, error rates and usability issues should be minimised for novice users	Time to complete a task / Number of tasks completed within a given time / Error rate / Error recovery time / Decision time
Learnability of product for novice users	Task structure and interaction style should be intuitive enabling users to reach acceptable performance levels within a specified time	Time to reach proficiency
Memorability of product for novice users	Task times and error rates should be minimised even after a period of non-use	Time to complete a task / Number of tasks completed within a given time / Error rate / Error recovery time / Decision time / Training time
Compatibility of product for all users	Task structure and interaction style should support the full range of user characteristics	Optimum number of commands / actions per task
Memorability of product for users over time	Task times and error rates should be minimised even after a period of non-use	Time to complete a task / Number of tasks completed within a given time / Error rate / Error recovery time / Decision time / Training time
Satisfaction of product short and long term	User satisfaction should remain high even after prolonged use	Positive versus negative statements from observations or questionnaires / opinions expressed during interviews
Satisfaction of product short and long term	User satisfaction should be high on first time use with all aspects of interaction	Positive versus negative statements from observations or questionnaires / opinions expressed during interviews
Perceived usefulness of product	Users should report a high likelihood of using the device during normal flight operations	Positive versus negative statements from observations or questionnaires / opinions expressed during interviews

FIGURE 4.2 KPI-associated metrics.

in Figure 4.2). However, it is important to note that these are intended to act as a starting point and should be tailored to meet individual project requirements. Evaluation against each KPI will however provide a means of assessment that can be used to give a complete usability evaluation of a product/system on the flight deck.

4.3 EXAMPLE: ASSESSING THE USABILITY OF TOUCHSCREEN TECHNOLOGY DURING TURBULENT CONDITIONS

The use of touchscreens on the flight deck has received a lot of attention over recent years (Harris, 2011; Stanton, Harvey et al, 2013). This is because they offer a new and intuitive way of interaction (Avsar et al, 2016). However, there is limited research available that assesses their utility during varied environmental conditions (e.g., turbulence). This contextual factor is however important because it may increase the likelihood of accidental touch, extended reach fatigue and operator difficulty in using

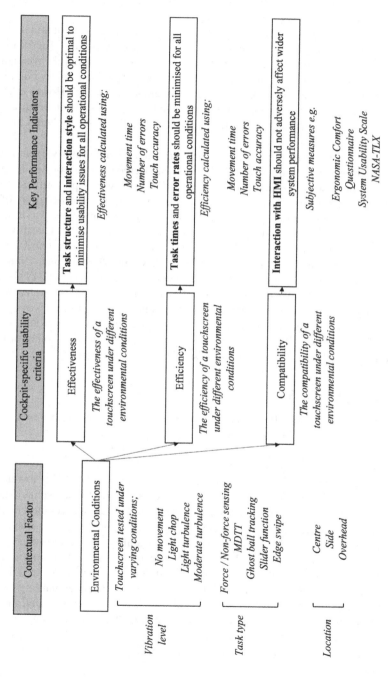

FIGURE 4.3 Investigative plan example (Coutts et al, 2019).

'soft' controls (FAA, 2011). Turbulence after all gives rise to inadvertent and unpredictable movements and therefore has the potential to impact the effectiveness, efficiency and compatibility of pilot interaction.

A recent study by Coutts et al. (2019) sought to explore the usability of touchscreens during different vibration conditions. The context-specific usability criteria identified in Figure 4.1 were used as a foundation in which a meaningful investigative plan could be developed (see Figure 4.3). They identified that the primary contextual factor relating to their investigations was 'environmental conditions'. However, it was recognised that 'environment' not only incorporated different levels of vibration, but also other variables including the location of controls on the flight deck and the type of task that may be completed.

The 'effectiveness' (in terms of task structure and interaction style) and 'efficiency' (in terms of task time and error rate) of touchscreen interactions under these varying conditions was assessed using different KPIs relating to performance (e.g., movement time, number of errors and touch accuracy). The 'compatibility' of the touchscreen (in terms of interaction with HMI) under varying conditions was assessed using subjective measures relating to comfort (e.g., Cornell University Questionnaire for Ergonomic Comfort; Hedge et al, 1999), user experience (e.g., System Usability Scale; Brooke, 1996) and workload (e.g., NASA-TLX; Hart & Staveland, 1988).

They found that the touchscreen was most effective for single-press tasks (i.e., tap or swipe), and that performance was most efficient for touchscreens located within the centre. However, the most compatible touchscreen location for users was the side screen. This early exploratory study provides an initial indication surrounding the usability of touchscreens in the cockpit. For further detail on the assessment of interface interaction under turbulent conditions see Chapter 9 (Section 4).

4.4 CONCLUSIONS

'Usability' is extremely difficult to define because it can mean different things depending upon context of use. This means that in order to assess the usability of a product/system, it is essential that context-specific usability criteria are developed with associated KPIs. As such, the purpose of this chapter was to present a set of cockpit-specific usability criteria that can be used to guide future usability evaluations within aircraft and flight deck operations. Comprehensive usability evaluations can be used to indicate the likely success of a product, lead to design recommendations to inform future designs, estimate any training requirements and also as a means to compare two or more similar products. The seven usability criteria presented in this chapter, along with their associated KPIs, may help deliver safer and more efficient aircraft operations. They therefore offer a compatible approach that can help demonstrate compliance to CS-25.1302 and further contribute to our understanding of what 'usability' means in the context of aircraft and flight deck operations. Defining and understanding usability and the context of use is one of the first important stages of user-centred design. The following chapter presents an overview of the user-centred design process across the phases of designing, modelling and evaluating new flight deck technologies.

5 The Design Process

5.1 INTRODUCTION

This chapter presents an overview of the design process that maps the stages involved in the development, design, modelling and evaluation of new concepts for integration into the modern cockpit. The design process has been developed in a way which to allow for user input to be included across all stages of new product development to facilitate user-led design. An overview of the requirements and processes involved in each of the design stages is presented before they are explored in further detail in the following chapters of the book.

5.2 THE DESIGN PROCESS

Changes to cockpit design can take considerable time to enact due to the lengthy processes involved in ensuring that they are safe and meet certification. It is important that future aircraft can benefit from technological advancements without being limited by time and expense but also, critically, that they uphold safety. A movement towards an 'open flight deck' aims to facilitate innovation within the cockpit to enable a platform that can undergo regular updates of flight deck applications. This will allow the development of new applications that can more effectively present information to pilots in the cockpit, as well as bring in new sources of information that may have not been previously viable. The application of Human Factors (HF) to the design, modelling and evaluation of new applications to this platform is critical to ensure safety and usability (Parnell et al, 2019).

An overview of this design process is shown in Figure 5.1, as well as the key actors that are required to facilitate and contribute to the work across each of the stages. The central key components of this process are 'design', 'model' and 'evaluate'. An initial 'generating requirements' process is also critical to provide the scope and boundaries of the intended concepts, from the outset. There is likely to be much complexity surrounding this process within the wider system of systems that comprises aviation, yet these four main components are considered invaluable within the design process.

DOI: 10.1201/9781003384465-5

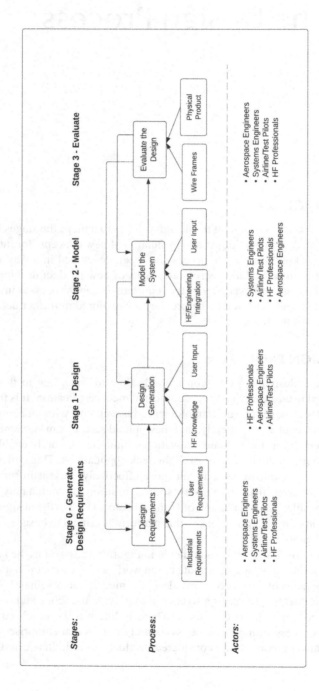

FIGURE 5.1   The design lifecycle for the development of flight deck technologies with the inclusion of HF practices and methodologies.

5.2.1 STAGE 0 – GENERATE DESIGN REQUIREMENTS

The development of design requirements is stated as stage 0, as it is important that the specific requirements are in place well before the designs and the design process itself are initiated. At this early stage it is important that a range of actors generate insight into their requirements of the technology. Any conflicting recommendations or visions of the intended design and use of the technology can then be overcome at this early stage, which is both cheaper and more effective. When working with an aerospace manufacturer their aerospace and systems engineers come together to identify the engineering requirements and functionalities of the future technology. These requirements can provide significant detail and may identify specific scenarios where they intend the functionality of the technology to provide novel and/or critical functioning, these are known as the 'concept of operations'. Once the initial requirements have been established, they form the template to generate designs from.

The presentation and the information surrounding the concept needs to be effectively communicated to all parties involved in the design process, including the HF practitioners and the pilots (airline and test) that will assist in the design creation. These parties must also have the opportunity to input into the requirements using their own expertise. Input from airline and test pilots is important to capture a range of experiences. As the airline pilots are the intended user base it is important that their perspective is incorporated in addition to test pilots who have different degrees of experience and backgrounds to those with commercial-based experience. Methods for generating requirements for these user populations include semi-structured interviews that enable qualitative and open-ended reports from the user, especially in relation to any predetermined scenarios suggested within the concept of operations. Interview data can then be mapped and modelled to determine the key areas of interest in relation to the proposed technologies. Further information on the methods involved at this stage, with case study examples, can be found in **Chapter 6.**

As shown in Figure 5.1, the process has interactive loops, and while the requirements feed into the following design stage they are also informed by other stages of the process. For instance, the development of the design may highlight further requirements that relate to how certain elements of the feature may look or function. Alternatively, it may highlight the inability to achieve certain requirements which will mean they need to be updated. Furthermore, the requirements should also act as a reference to determine the success of the design at the end of the process, ensuring that the developed concept captures all the functions that were required.

5.2.2 STAGE 1 – DESIGN

Once the design requirements have been established by all actors involved including the engineers, Human Factors practitioners and the pilots, the initial design stage begins. During the early stages of the design process multiple different design concepts can be generated in a cost-effective way through drawings and simple digital mock-ups. Whilst there a number of HF methods available to assist designers throughout the design lifecycle, we found the Design with Intent (DwI) method to be particularly effective within this context. DwI was originally developed by Lockton et al (2010) following the recognition that designers lack guidance on choosing appropriate design techniques to influence end-user interaction. The DwI toolkit is marketed as a 'suggestion tool' that can be used to promote directed brainstorming. Conducting workshops with both aviation engineers and commercial pilots and using Human Factors practitioners as moderators can facilitate hugely informative discussions on the design requirements and their actualisations from different perspectives (Parnell et al, 2019). **Chapter 7** presents further information on the DwI methodology with examples of the designs that were generated from these workshops.

5.2.3   STAGE 2 – MODEL

Once the initial design concept(s) have been generated the designs are then modelled to understand how they will function within the wider system that they will operate within, including their integrations with other functionalities and environmental factors. There are a multitude of different Human Factors modelling techniques available to review different facets of the design. Error modelling has long been a key element within HF approaches. However, there are calls within the HF domain to move away from the term 'human error' (Dekker, 2011; Read et al, 2021; Salmon et al, 2017), with the suggestion that it is slowing the pace of safety improvement (Read et al, 2021). Instead, a movement towards a systems perspective is advocated which takes the system as the unit of analysis, rather than its comprising elements. These analyses should review the conditions and interaction between components in their normal functioning to understand how variability in system performance may lead to system failure (Dekker, 2011; Hollnagel, 2014; Read et al, 2021; Salmon et al, 2016). This means not relying on the benefit of hindsight but proactively reviewing system functioning to support any positive variability and reduce negative variability (Read et al, 2021).

The Systemic Human Error Reduction and Prediction Approach (SHERPA; Embrey, 1986) has been a prevalent error prediction method. Originally designed with a focus on 'human error', the analysis reviews all tasks comprising an activity and identifies their opportunity for 'error' against the SHERPA taxonomy (Embrey, 1986). Developments of this method (e.g., Stanton, 2004) have allowed it to move beyond a simplistic 'error' analysis to identify how the interaction between different tasks and actors can facilitate error recovery, as well as where system recovery is not possible, and failures arise. Reviewing these 'errors' as potential system failures also allows for resilience engineering interventions to be identified and proposed (Parnell et al, 2023). Further details on this method and its application can be found in **Chapter 7**.

Alternative HF modelling techniques can model the wider system within which a technology will be placed, such as operator event sequence diagrams (OESD; Brooks, 1960; Kurke, 1961). OESDs are a useful tool in mapping human–automation interactions (Banks et al, 2014; Harris et al, 2015; Stanton et al, 2022a;) with validity (Stanton et al, 2022b). They present the allocation of functions to human actors and non-human actors equally, to show the distributed performance of the systems in achieving its goal (Hutchins, 1995; Stanton et al, 2022b). This method is presented in **Chapter 8** with a case study example.

Psychology modelling techniques can also be useful when assessing the impacts that new technological designs will have on the human operator. A core aspect of the case studies that we discuss in this book incorporates human decision-making, with the designs providing information to the pilot that requires them to make informed and justifiable decisions, some of which may be safety critical. It is therefore useful to understand the core theories surrounding human decision-making, and in particular the field of Naturalistic Decision-Making (NDM). NDM is the study of decision-making in association with the environment that the decision occurs and the decision maker themselves (Klein, 2008). The field of NDM diverged from previous decision-making research at the time of its conception, which had assessed controlled, structured environments where the decision-maker was passive to the outcome (Klein, 2008). The inception of the field of NDM began with reviewing decision-making 'in the field' to understand the strategies behind natural decisions (Klein, 2008). Multiple different theories of NDM have since come to fruition (Lipshitz, 1993) and there is still much debate in the field over the best ways to conceptualise decision-making under naturalistic conditions (Lipshitz, 1993; Lipshitz et al, 2001; Lintern, 2010; Naikar, 2010). In **Chapter 8** we present three key decision-making models that are useful to the aviation domain and the study of pilot decision-making; the Recognition Primed Decision Model (RPDM; Klein, 1989), Decision Ladders (Rasmussen, 1983) and the Perceptual Cycle Model (PCM; Neisser, 1976).

5.2.4   STAGE 3 – EVALUATE

Once designs have been generated and then modelled to determine their feasibility, they require evaluation (Stanton et al, 2013, 2014). This occurs across two stages. Initial wire-frame evaluation involves drawings of the design in its basic form, which is time and cost saving. Multiple designs can be initially evaluated and compared at the wire frame stage. We use heuristic-based evaluations on the wire frames to allow for down selection. Conducting a heuristic evaluation is a cost-saving and relatively quick and easy method for evaluating multiple designs before they go into production. The evaluations can be done on rudimentary mock-ups of the displays such as PowerPoint presentations or simple drawings. Heuristics provide generic usability criteria to assess the interfaces against. Neilsen's 10 Usability Heuristics (Neilsen, 1994) are the most widely used and accepted heuristics for interface evaluation and so were applied within this work.

Heuristic evaluations can be conducted with Human Factors professionals, who have an understanding of interface usability, as well as the intended user group who have an understanding for the context in which the interface is to be used. Evaluations from both groups can be complimentary and enhance the validity of the findings. The most successful designs are then selected for physical reproduction and integration in a flight simulator with a replicated flight deck for physical testing.

Reliable and robust evaluation occurring in this second stage is required to ensure that the designs are usable and safe before they are considered for integration into future flight decks. This requires a realistic flight simulator and user testing with the intended user base, within our work this was commercial airline pilots of ranging experiences and backgrounds. User testing must be designed to include a baseline condition and allow sufficient power to be drawn from the data to generate conclusions and enable statistical significance to be obtained. Further information on simulator evaluation is presented in **Chapter 9.**

It is unlikely that the process of designing new technology will go through these three stages in a single linear fashion. The inclusion of multiple feedback loops across these stages (as indicated in Figure 5.1) is intended to enhance the final design and allow for refinement in response to challenges that may arise. The process must be iterative to allow the best chance of success. The perspectives from all actors involved should be recognised and collaborative efforts across this process will enable the best results.

5.3   CONCLUSION

Collaboration between aerospace manufactures and engineers, systems engineers, pilots and HF practitioners has sought to develop an HF design process which can be utilised within the early stages of interface design. Following the process, and its feedback loops, has been valuable in generating informative discussions between different actors who hold different perspectives on the design of aviation technologies. Sometimes these may be conflicting, yet these methods show these conflicts may be identified and challenged early in the design process, before they become larger issues later on. This process shows how HF practices can, and should, be integrated into the design process from the very start. Inclusion of the end-user can generate huge benefits to the development of usable designs that can be integrated alongside their current practices. This chapter has provided an initial overview of the design process and the purpose of each of the stages. The following chapters of this book will provide more insight into the methods used within each stage, as well as applying them to real-world case study examples.

6 Design Requirements

6.1 INTRODUCTION

This chapter will detail what is required during the initial stage of the design process; defining new technological concepts and outlining their requirements. This is an important starting point as it provides an understanding of the context of use that technologies will need to perform within. It will also recognise the interrelating agents and factors surrounding the integration of new technologies within operating systems.

The reasons for designing and developing new technologies are numerous. Here we will focus on the need to advance current systems by developing new technology that is integrated within current flight deck devices, as well as designing completely new technologies that need to be defined from the very beginning of the design process. Advancing current systems is motivated by a need to overcome possible failings in the current system such as incidents and accidents or to adapt to changing and advancing environments. Technological advancement can also facilitate novel features and applications that have not been previously available. The two motivations for new technological design will be discussed in this chapter with relevant examples.

6.1.1 ADVANCING CURRENT SYSTEMS

As technologies become more advanced they bring the opportunity to alter how work practices are performed. Increased levels of autonomy may change what is required of the human operator and the interactions between the user and the technology. Within the process of redesigning work practices it is vital that the processes involved in the original systems are fully captured and used to inform redesign. The requirements for the future systems can then be overlaid onto the current practices to ensure that any new potential for error or unwanted interactions do not occur.

A case study example of redesigning the current aircraft Flight Management System (FMS) to utilise a touchscreen interface instead of traditional physical buttons is presented to outline this process.

DOI: 10.1201/9781003384465-6

6.1.2 CASE STUDY EXAMPLE: ON-BOARD FLIGHT MANAGEMENT SYSTEM

Before take-off, pilots of commercial aircraft are required to enter flight information into an FMS. The FMS is a multi-purpose technology that holds navigation, performance and aircraft operation information. It aims to provide virtual data and operational harmony across the entirety of the flight, from take-off to landing and engine shut down. It has some autonomy and can communicate with other components of the flight deck including the Engine Indicating and Crew Alerting System (EICAS) and Electronic Centralized Aircraft Monitor (ECAM). The interface to the pilot is the control display unity (CDU). This has a full alphanumeric keypad and a liquid crystal display (LCD) (see image in Figure 6.1).

The Flight Management Computer (FMC) is another central component of the FMS which stores a database containing a large number of flight plans that have predetermined operational parameters, selected by the pilots.

FIGURE 6.1   Image of CDU display.

The buttons next to the screen are used to select the menu options. The alphanumeric keypad is used to input data. The other buttons relate to specific operations and information.

The pre-flight setup has traditionally been viewed as an arduous process requiring pilots to manually input large volumes of data in the FMS using the CDU, with limited usability. As systems have advanced, so too has the process of system initialisation. Nowadays, pilots have access to pre-programmed company routes and datalink services.

With advances in technology, there was a proposed redesign of the FMS to enhance usability and functionality. A new FMS application sought to revolutionise the way in which the FMS is programmed by integrating numerous functionalities onto a single touchscreen display. The use of touchscreen technology facilitates more usable interactions in comparison to the contemporary system pictured in Figure 6.1 which requires users to select an input from a list and then select the location on the CDU that they would like to enter it in to.

When re-designing the FMS for a modern touchscreen interface it is important to consider its functionalities and uses to ensure that these are not lost in a future system. Furthermore, they can be used to suggest ways in which task–user interaction may be improved. To understand the functionality of the system, a Hierarchical Task Analysis (HTA) was conducted to map out the tasks involved in FMS interaction. This was applied to two specific use cases: manual data entry during the system initialisation process (i.e., route entry) and entering a diversion airport during flight.

6.1.2.1 Hierarchical Task Analysis

Hierarchical Task Analysis (HTA) is a task analysis method that is used to provide an exhaustive description of tasks in a hierarchical structure of goals, sub-goals, operations and plans. It is now considered to be one of the most popular and widely used HF techniques due to its flexibility and scope for further analysis (Stanton et al. 2013). See Chapter 3 (Section 3.4.3) for further information on this method.

The process of constructing the HTA for the FMS use case scenarios included attending two FMS training workshops. These training sessions were hosted by an aircraft manufacturer and conducted by an experienced FMS user who trains pilots in CDU interaction. Information from the training session was used to develop initial HTA representations of the two case studies. These were then shared with the experienced trainer for review. They were also reviewed by two commercial pilots (who acted as subject matter experts; SMEs) to ensure that the models were accurate. This was an iterative process with multiple updates occurring due to the various ways in which pre-flight setup may be achieved and performed by individual pilots. This was concluded once a consensus was reached.

The HTA representations were developed using the method outlined by Huddlestone and Stanton (2016) which allows for repetitive processes to be easily captured and basic functions to be clearly identified. The final HTA relating to pre-flight setup is presented in Figure 6.2. It outlines three main subtasks (enter current position; set up route; and check for discontinuities).

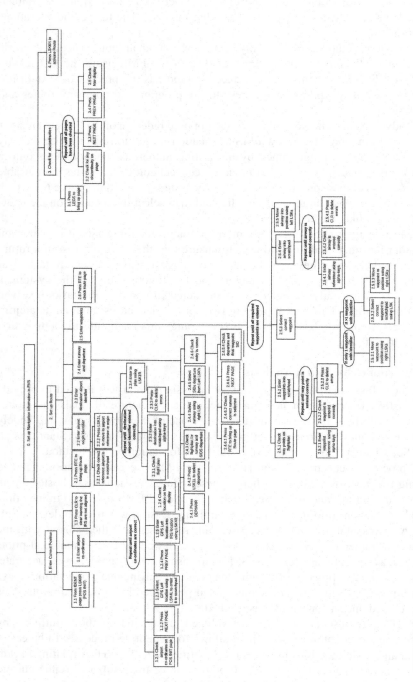

FIGURE 6.2 Current practice in relation to pre-flight set up in the FMS.

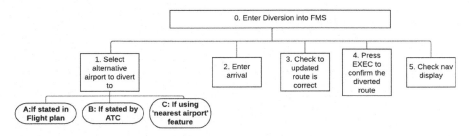

FIGURE 6.3 Top-level tasks for the function 'enter diversion into FMS' with the three options for selecting an alternative diversion airport.

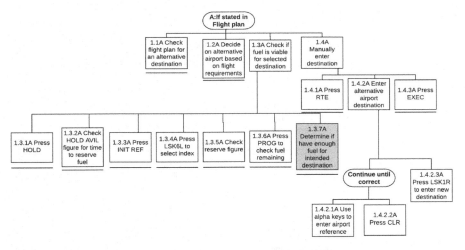

FIGURE 6.4 HTA for the tasks involved in option A for selecting a diversion: 'if stated in the flight plan'.

The second scenario is related to an in-flight route change (i.e., entering a diversion airport). The final iteration of the HTA relating to this specific scenario is presented in Figure 6.3. It highlights that there are three different ways in which an alternative airport may be entered into the FMS;

A) From the flight plan (Figure 6.4)
B) From air traffic control (ATC) (Figure 6.5)
C) Using the 'nearest airport' feature (Figure 6.6)

Now that the HTA has mapped out all the tasks and sub-tasks involved in the FMS data entry processes, the possibility for errors to occur within the interactions can be reviewed. There are numerous different error modelling and identification techniques that can be used. Further information on the different methods can be found in Stanton et al (2013). The method we choose is the Systematic Human Error Reduction and

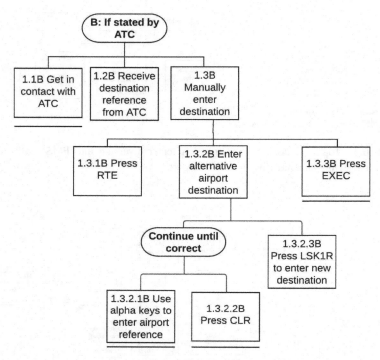

FIGURE 6.5 HTA for the tasks involved in option B for selecting a diversion: 'if stated by ATC'.

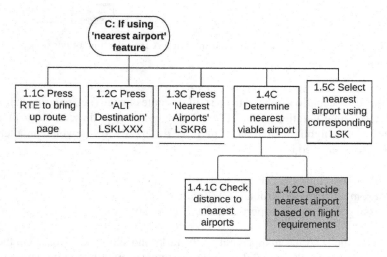

FIGURE 6.6 HTA for the tasks involved in option C for selecting a diversion: 'if using nearest airport feature'.

Prediction Approach (SHERPA) (Embrey, 1986; Stanton et al, 2013). This is because SHERPA has been heralded as the most suitable Human Error Identification (HEI) technique for aviation in contrast to other HF techniques (Salmon et al, 2002). The guidewords used in the accompanying SHERPA error taxonomy (Embrey, 1986) permit the identification of errors most likely to be made by a pilot (e.g., 'A6 – Right action on wrong object' corresponds to a pilot incorrectly pressing a button). Other error taxonomies do not allow analysts to extrapolate to the same degree. Chapter 3 (Section 3.4.4) provides some more information on this method.

6.1.2.2   Using SHERPA to Understand and Mitigate Human Error when Interacting with the FMS

The bottom-level tasks identified within the HTA (Figures 6.3–6.6) were used as the main input to the analysis (Huddlestone & Stanton, 2016). Errors that could occur for each of these tasks were then categorised against the SHERPA error taxonomy (Embrey, 1986; Stanton et al, 2013; see Chapter 3, Table 3.2). The SHERPA outputs from the use cases reflecting the pre-flight set up and entering a diversion airport are now presented.

6.1.2.2.1   Pre-flight Setup

For each bottom-level task within the HTA, a comprehensive assessment was made to determine error consequence, identify possible recovery routes and assess the likelihood and criticality of its occurrence. Ideas for remedial measures were also noted as part of the SHERPA methodology (Stanton et al, 2013).

From this analysis, a total of 89 possible errors were identified (see Table 6.1). The most common error category was action errors (with a total of 52). The most common specific error type was action error A6: Right operation on wrong object (37 errors). This simply reflects the potential number of times pilots may press the wrong button on the keypad. The second most common error category was related to checking errors. These errors may signal issues of complacency in which pilots fail to check data entered into the FMS.

Each error was then assessed on its likelihood and criticality using the following definitions;

- Likelihood of error: Low (never/rarely), Medium (occasionally/has happened before), High (frequent)
- Criticality of error: Low (no danger to life or injury), Medium (risk of injury), High (possible loss of life)

The majority of errors were determined to have low likelihood and low criticality (see Table 6.2). Whilst the analysis does not highlight any urgent cases for design improvement, the FMS could be improved by incorporating larger, modern displays, with bigger buttons to limit the occurrence of incorrect button presses. In addition, many of the incorrect button presses may be remedied through the addition of a CLR ('clear') button.

TABLE 6.1
Frequency of Error Types for Pre-flight Setup

Error categories	Error sub-categories	Errors (*n*)
Action	Right operation on wrong object (A6)	37
	Action omitted (A8)	15
	Total Action Errors	*52*
Checking	Check omitted (C1)	14
	Check incomplete (C2)	5
	Right check on the wrong object (C3)	1
	Total Checking Errors	*20*
Selection	Selection omitted (S1)	3
	Wrong selection made (S2)	3
	Total Selection Errors	*6*
Communication	Information not communicated (I1)	2
	Wrong information communicated (I2)	2
	Total Communication Errors	*4*
Retrieval	Wrong information obtained (R2)	7
	Total Retrieval Errors	*7*
Cumulative total		**89**

TABLE 6.2
The Frequency of Error Likelihood and Criticality Ratings for Pre-flight Setup

	Likelihood		
Criticality	Low	Medium	High
Low	86	3	0
Medium	0	0	0
High	0	0	0

6.1.2.2.2 Entering a Diversion Airport

For each of the different ways in which a diversion airport may be entered, the type of errors that may be made were reviewed with the SHERPA error taxonomy. From Table 6.3, it is clear that method A (from the flight plan) has the potential to yield the highest number of errors. Specifically, action errors are the most prevalent within this case – as pilots will be required to manually input the necessary diversion airport data. Method B (from air traffic control) and C (using the nearest airport feature) scored equally overall, although there were slightly more action errors associated with method B, but method B had less checking errors than method C.

Each error was again assessed on its likelihood and criticality. The majority of errors were determined to have low likelihood and low criticality (see Table 6.4). The errors that yielded low likelihood and medium criticality ratings were related to the

TABLE 6.3
Frequency of Error Types for Entering a Diversion Airport

Error categories	Error sub-categories	Method A	B	C
Action	Right operation on wrong object (A6)	23	18	17
	Action omitted (A8)	5	5	4
	Total Action Errors	*28*	*23*	*21*
Checking	Check omitted (C1)	7	4	5
	Check incomplete (C2)	0	0	1
	Right check on the wrong object (C3)	3	1	1
	Total Checking Errors	*10*	*5*	*7*
Selection	Selection omitted (S1)	0	0	0
	Wrong selection made (S2)	2	2	2
	Total Selection Errors	*2*	*2*	*2*
Communication	Information not communicated (I1)	0	0	0
	Wrong information communicated (I2)	0	0	0
	Total Communication Errors	*0*	*0*	*0*
Retrieval	Information not obtained (R1)	1	0	0
	Wrong information obtained (R2)	6	2	2
	Total Retrieval Errors	*7*	*2*	*2*
Total		47	32	32

TABLE 6.4
The Frequency of Error Likelihood and Criticality Ratings for Entering a Diversion Airport

Criticality	Likelihood Low	Medium	High
Low	45	2	0
Medium	10	0	0
High	0	0	0

possibility that the flight crew could make the wrong divert decision. Errors yielding medium likelihood, low criticality ratings were related to using incorrect alpha key presses (similar to that in the pre-flight setup).

From the identification of the errors and the classification, remedial measures were proposed. These included the following, which were proposed for further exploration with the aerospace manufacturer:

• Utilisation of a touchscreen interface;
• Force feedback for button press inputs;

- Visible changes to the display in response to inputs;
- Highlighting important areas of the display;
- Reducing the number of pages to prevent the requirement for continuous tracking between screens (i.e., present all relevant information together);
- Incorporate a back button or CLR button; and
- Link FMS directly to the flight plan (i.e., taking advantage of increasing levels of connectivity).

The FMS is however a highly complex system and much more work is required to understand how changes to one aspect of the system may impact upon the operation of the wider system. Yet, it does seem likely that greater levels of connectivity could significantly reduce the level of interaction that is currently required between the flight crew and the FMS. For instance, route data may be preloaded into the FMS, requiring pilots to check data against their flight plans. Action errors therefore could be greatly reduced, but this brings with it a risk of increasing the number of checking errors.

Notably, both the HTA and SHERPA analysis have been used to support the development of a Concept of Operations document (CONOPs) for future FMS interactions. Going forward, we need to think more about what the graphical FMS application may look like and how it may be integrated onto the flight deck.

N.B. Further work on this FMS case study example is explored in Chapter 8 (Section 8.2.1). Further applications of the SHERPA methodology are presented in Chapter 7 (Section 7.4) to assess the recommendation that the methodology makes to future design recommendations.

6.2 DEFINING NEW CONCEPTS

This section will detail how the requirement for new product design can be understood through reflecting on the efficacy of current systems, to advance their functionality and usability. Critically this involves understanding how the users interact with the technology and the environments within which it is used. Taking a user-centred approach to this is valuable in understanding the user's requirements for new technology. A case study example of an engine oil starvation alert system, which was under design by an aerospace manufacturer, is used here to illustrate how user requirements are generated.

6.2.1 CASE STUDY: OIL STARVATION ALERT

Aircraft engine oil leaks are a rare occurrence, yet they do pose a significant threat to the safety of the aircraft and the way in which they are managed is important to maintaining both the safety of those on-board and efficiency of the airline. Currently, the pilot is only alerted to a low oil pressure warning once pressure levels have reached minimum limits. This reduces the options that they have available to preserve and maintain the safety of the flight. The pilot must determine the validity of the message and then assess the severity of a suspected oil leak before taking the necessary action.

Although rare, it has been noted to occur in the past and can pose a serious threat to safety if it is not dealt with appropriately (Australian Transport Safety Bureau; ATSB, 2012, 2017). The seriousness of the event is further indicated by the low oil pressure warning scenarios that are included in the simulator training that pilots must regularly undergo.

The response of pilots who have been exposed to oil leak scenarios in real life has been to either throttle the engine back to preserve oil, or shut the engine down completely and divert the flight (ATSB, 2012, 2017). Engine shutdown events and diverted flights have obvious repercussions on the flight operator and those involved in the maintenance of the aircraft, as well as possible loss of life in the most extreme circumstances. It is therefore in the interest of flight operators to prevent these incidents from occurring.

The presentation of accurate, reliable and up-to-date information to the pilot on the status of the engine is one way in which such incidents can be prevented. The current process of supplying the pilot with information only after low levels of oil have been reached may not be the most proactive approach in allowing effective mitigation strategies to be put in place. Pilots have to do extra work to first ensure that the information is accurate and then calculate the options available to them at that moment in time. Work presented in this chapter suggests that the development of an engine monitoring system could provide the pilot with more precise and timely information on the status of the engine, with data that could be interpreted to indicate an oil leak at an earlier and more opportune moment than is currently available. Furthermore, with more forewarning they could also make better informed decisions on what options they have available to them.

Future technology hopes to provide enhanced information to the pilot on the status of the oil leak and provide a decision support tool to aid the pilot in taking appropriate action before it adversely affects the flight or the aircraft.

6.2.1.1 User Requirements

Often the user can be overlooked within the early stages of the design process (Gould & Lewis, 1985; Stanton & Young, 2003). Yet, gaining an understanding of their needs, desires and the plethora of tasks that they are required to undertake can provide useful insights into the requirements of future systems (Banks et al, 2020a; Parnell et al, 2021b). Therefore, an understanding of how an engine condition monitoring tool can align with the current processes that occur during engine failure events is required. The pilot's response and subsequent decision-making to manage the situation can be reviewed to determine how an engine monitoring tool could assist decision-making and facilitate improved awareness of the state of the engine after sustaining a bird-strike during take-off. Central to this are the interactions between the different elements that comprise the system of systems that is vital to understanding complexities in the aviation domain (Harris & Stanton, 2010).

Pilot interactions and/or display indicators cannot themselves be considered in isolation, they must be considered within the broader sociotechnical system that exists in the cockpit (Plant & Stanton, 2012). We used the Perceptual Cycle Method (PCM;

Neisser, 1976) to capture the interactional nature of the aviation system in critical situations (Plant & Stanton, 2012).

6.2.1.2 The Perceptual Cycle Model

The PCM can account for the interactional nature of the environment and wider system (Stanton et al, 2010) whilst also capturing the accounts of individuals and their cognitive processing (Plant & Stanton, 2012). The model is underpinned by Schema Theory (Bartlett, 1932). Schemata are knowledge clusters that are structured upon experiences that are similar in nature and capture commonalities that represent the experience. They provide mental templates that can inform future behaviours, as well as being fluid to updating upon exposure to new experiences. They can also allow abstract behaviour and knowledge to be assimilated in order to determine an appropriate response. The key components of the PCM are 'Schema', 'World' and 'Action' (Neisser, 1976). The key premise is that an individual's interaction with the world and their internal thought processes are reciprocal and influence each other in a cyclical manner. As can be seen in Figure 6.7, the cyclic behaviour within the model can be bottom-up (BU) or top-down (TD). Schema are initially triggered from the world, and information available within it, via a bottom-up process. The schema that relate to expectations and/or past experiences of the information that they are sampling in the world is then activated. Top-down processing then occurs whereby actions are activated in line with the processing of the schemata, to respond to the event in the world.

Application of the PCM across ergonomics and safety-critical domains has informed system theorems (e.g., Smith & Hancock, 1995; Stanton et al, 2006). Application of the PCM to the aviation domain has shown that it can account for erroneous events through providing a wider, systems viewpoint on why events may

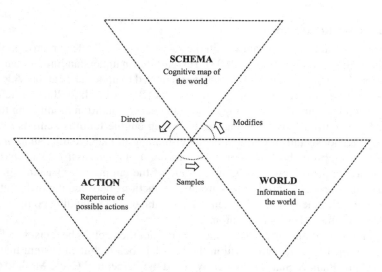

FIGURE 6.7   Representation of the PCM adapted from Plant and Stanton (2012).

have made sense at the time (Plant & Stanton, 2012. 2013). Plant and Stanton (2012) demonstrated how the PCM showed how the expectations of the pilots, based on a previous generation of aircraft, led them to shut down the wrong engine when an engine fire was discovered on a British Midland flight that was diverted to East Midlands airport in Kegworth. Had the pilots applied the same decision-making on the aircraft that they had more experience with, the aircraft would not have crashed. The PCM has also been shown to present options to prevent safety adverse events from happening again in the future (Revell et al, 2020).

By modelling the current processes involved in responding to and managing engine failure events, the PCM can be used as a starting point to then see if, and what, modifications may be useful in the development of engine monitoring technologies. The engine monitoring tool would need to fit within the wider system of the flight-deck and not compromise any other processes that may be occurring in unanticipated events. Therefore, capturing the perceptual processing, including the interactions with the wider environment that pilots currently perform, is vitally important for the future design of the system. Conducting interviews with active airline pilots to populate the PCM will allow the intended user population for the engine monitoring tool to be integrated into the design process at the early stage.

The Schema World Action Research Method (SWARM) is an interview methodology that was specifically developed to understand aeronautical critical decision-making in relation to the PCM (Plant & Stanton, 2016); see Chapter 3 (Section 3.4.1.1) for more information. This interview method is applied here.

6.2.1.3   Method for Eliciting User Requirements using the PCM

6.2.1.3.1   Participants
Six commercial airline pilots were recruited to take part in this study (two female, four male), aged between 26 and 35 years ($M = 30.17$, $SD = 3.02$). All participants were qualified fixed-wing ATPL or CPL pilots with an average 3692 hours flight experience ($SD = 570.39$) and had held their licences for an average 8.08 years ($SD = 1.59$). Interviews lasted for approximately 1.5 hours and participants were reimbursed for travel and time spent participating in the study. The study was ethically approved by the research institutes Ethical Research Governance Office (ERGO; reference ID: 40619).

6.2.1.3.2   Procedure
The participants were presented with a hypothetical scenario relating to a suspected aircraft engine oil leak. This stated the following:

> 'An incomplete maintenance action has resulted in an oil leak in the aircraft engine. You are in cruise'.

Following the presentation of the scenario, participants were invited to take part in a semi-structured interview using exemplar prompts from the SWARM repository. To get their initial thoughts on the scenario and their response they were asked to give a brief initial overview of their thoughts in relation towards the scenario, prompted by two questions:

1. Currently, how would you be informed of an oil leak in an engine?
2. How would you respond to it?

While these questions were not part of the SWARM interview prompts, they allowed the participants to speak openly about their perception of the scenario as well as ask any further details from the researchers. After this the SWARM prompts began.

A down-selected set of SWARM prompts were selected from the original 95. Plant and Stanton (2016) suggest that not all SWARM prompts are relevant to all events and so down selection allows the interview to be more efficient. Plant & Stanton (2016) state that preferably this would use the top five subtypes for each PCM category. The SWARM prompts were reviewed by Human Factors experts and 37 were selected. Example relevant prompts included *'what would you be looking at on the technological system during the scenario?'*; *'would you require information from others?'*; *'what information would you use to assess the severity of the problem?'*. Participants were asked to speak openly and honestly and in as much depth as possible. This enabled a full account of their anticipated response to the scenario, capturing their schema representations, information they would access from the world and the actions they would be taking.

6.2.1.3.3   Data Analysis

All interviews were audio recorded and subsequently transcribed by the researchers. The transcripts were then coded to the Schema, Action or World feature of the PCM. Please see Plant and Stanton (2013, 2016) for further details on this process. The process of mapping the interview reports on to the PCMs was conducted in an iterative manner until the researchers were satisfied that the mapped decision-making process accurately reflected the interview data. The data were used to develop an amalgamated representation of the pilots' PCM for the given scenarios. As pilots are highly trained and the procedures that they conduct on the flight deck are heavily standardised and regulated, the reports were largely similar. The generated PCM containing the combined responses of the pilots was then reviewed by an independent subject matter expert, with over 10 years flight experience and a strong background within the Human Factors discipline to ensure accuracy.

6.2.1.3.4   PCM of Current Processing in Response to an Aircraft Engine Oil Leak

The final amalgamated PCM of current practice is presented in Figure 6.8 and outlines the perceptual cycle processes of a pilot when they are subjected to dealing with a suspected engine oil leak. Information available in the 'world' acts as the impetus for diagnosis and option generation. To first diagnose and understand the problem presented to them, pilots would attend to any warning messages on the Electronic Centralised Aircraft Monitor (ECAM) system, also sometimes referred to by some manufacturers as the Engine Indicating and Crew Alerting System (EICAS). This system provides data to pilots on the status of a variety of aircraft systems, as well as providing warnings and alerts when parameters reach unusual levels. These warnings are colour-coded to differentiate their urgency for attention and action by the pilot.

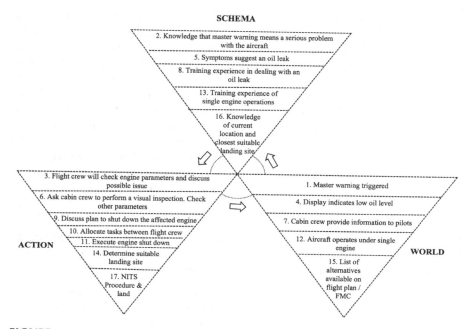

FIGURE 6.8 Amalgamated PCM of pilots' current approach to dealing with an engine oil leak.

Participants reported that when receiving an ECAM/EICAS message in this scenario they would seek to validate it by checking the most appropriate system parameters, in this case oil pressure, oil quantity and oil temperature. All of these artefacts are available in the 'world' via the Engine Display. Appraisal of these dials would enable the pilots to determine whether the warning is genuine or spurious. Particularly on older aircraft, sensors are vulnerable to exhibiting false indications and therefore the presence of a single abnormal reading may not be a valid indication of failure. Thus, to confirm a genuine failure, flight crews must inspect multiple sources of information. Five participants suggested they might also seek assistance from the cabin crew in terms of performing a visual inspection of the engines. This would provide an opportunity to gather more 'world'-based information that would be used to trigger relevant 'schema', which was their previous experience in dealing with available information. This process shows how information in the 'world', that is the presence of a master warning, can go on to trigger relevant 'schema'. For example, the presence of the master warning suggests something is seriously wrong, leading to subsequent 'action' which includes checking affected engine system parameters.

When it came to making a decision, all six pilots spoke about the utilisation of the 'DODAR' decision aid (Diagnose, Options, Decision, Assign task, Review) or variations of such, for example, T-DODAR (Time, Diagnose, Options, Decision, Assign task, Review) in order to help them systematically reach a decision (Walters, 2002). Different airlines have alternative aids but each has a similar theme to help pilots with decision-making. Each element of the tool details key requirements and the order dictates the order the requirements should be fulfilled. 'Diagnosis' is the requirement to gather as much information as possible to determine and confirm the problem. 'Options' required the pilot(s) to consider all possible alternatives relating to all possible actions. 'Decide' requires the pilot(s) to come to the most appropriate selection of action(s). Once the decision is made, they must then 'Assign tasks' to allocate tasks between the pilot flying and pilot monitoring. The 'Review' aspect then denotes that the resulting consequences and emerging situation should be continuously reviewed to ensure the desired outcome is achieved. It was evident from the interviews that the DODAR aid was central to the decision-making process. The detail they gave regarding this linked into the SWARM interviews to allow participants to discuss the factors that guided their decisions and the resulting actions.

6.2.1.3.5 Generating Design Requirements

The second part of the interview asked participants to consider how a future Engine Monitoring Assistant (EMA) system may influence their responses to the previous scenario. Here the updated scenario stated the following:

> 'An incomplete maintenance action has resulted in an oil leak. An automated system detects a non-normal change in system parameters and notifies you of a non-critical oil leak (i.e. sufficient levels to complete flight to intended destination safely).'

Participants were again asked key SWARM prompts of relevance to generate data that could also be used to inform the development of a PCM that could inform how participants thought a future system could assist them. From the PCMs and the transcriptions, the researchers reviewed how the pilots could be further supported with their decision-making to inform the design recommendations. Importantly, participants were not given specific details of how this system may work, rather they were asked to determine how they may want it to work and what utility it may have.

All six pilots recognised that an early warning from an EMA system would indicate that minimum thresholds had not yet been met. For example, one pilot stated: '*Time is everything in an aircraft, with failures, with everything. It buys you time so you can sit, think, discuss with the company, come up with a plan and have a bit more time before it is critical*' (Participant 2). Participants stated that they would still utilise the DODAR decision-making tool. This is an airline-regulated decision tool that pilots must use to guide their decision-making and therefore it would still be applied when the EMA is present. In this scenario, the flight crew would begin monitoring and trending oil in order to rule out the potential for spurious sensor readings as soon as the assistant system alerted them to a possible issue. With engine parameters continuing to show an abnormal downward trend, the flight crew would use their prior experience (both through operational activities and training) to determine that an oil leak was indeed occurring. In terms of a pilot's PCM, Figure 6.9 demonstrates that information in the world (i.e., the system triggers alert in relation to the oil system) can trigger underlying schemata (i.e., EMA provides an advisory warning that may be wrong and therefore requires further investigation), which then goes on to influence

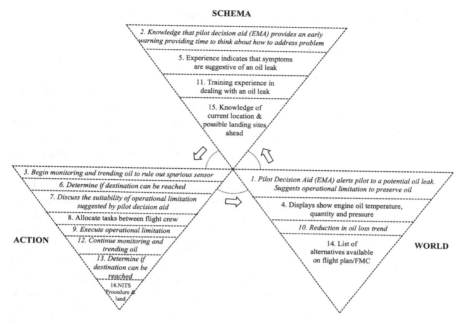

SCHEMA

2. Knowledge that pilot decision aid (EMA) provides an early warning providing time to think about how to address problem

5. Experience indicates that symptoms are suggestive of an oil leak

11. Training experience in dealing with an oil leak

15. Knowledge of current location & possible landing sites ahead

3. Begin monitoring and trending oil to rule out spurious sensor

1. Pilot Decision Aid (EMA) alerts pilot to a potential oil leak. Suggests operational limitation to preserve oil

6. Determine if destination can be reached

7. Discuss the suitability of operational limitation suggested by pilot decision aid

4. Displays show engine oil temperature, quantity and pressure

8. Allocate tasks between flight crew

9. Execute operational limitation

10. Reduction in oil loss trend

ACTION   12. Continue monitoring and trending oil

14. List of alternatives available on flight plan/FMC   **WORLD**

13. Determine if destination can be reached

16. NITS Procedure & land

FIGURE 6.9 Amalgamated PCM of pilots' future proposed approach to dealing with an engine oil leak with an automated system assistance.

action (i.e., monitor and trend oil system parameters to confirm what the problem may be). Importantly, the majority of the PCM remains similar to that presented in Figure 6.8 (i.e., pilots would utilise the same information in the world which would go on to trigger the same underlying schema). As one pilot stated: '*I would do the same checks. I would have a look for myself...*' (Participant 4). However, this is likely to be a product of the strict training a pilot goes through. For example; '*...I think that is something that pilots always do. We have to check, check and double check*' (Participant 1). With this in mind, it is mainly the 'Action' phase of the PCM that changes as a result of EMA implementation which is what we would expect.

Whilst in the first scenario (current practice) pilots would choose to shut down the engine, the general consensus amongst the pilots for the second scenario (with the EMA) would be to accept the information provided by the pilot decision aid and therefore execute the operational limitation. Pilots would then re-evaluate whether they would still be able to reach their destination, based upon their monitoring and trending of oil quantities. A key difference between the current processes in Figure 6.8 and anticipated processes in Figure 6.9 is that pilots are better informed with the future system on the remaining level of oil and if they can reach their destination. Pilots suggested that if they could receive more up-to-date information regarding the oil level in the case of a leak then there would be less ambiguity and, therefore, they could make more informed decisions about shutting the engine down. Importantly, pilots did not think that the addition of an EMA would slow down a pilot's response to an incident. Instead, the provision of an early warning would enable pilot responses to be more considered. The process of diagnosis, the options that are available, the allocation of tasks and key decision points remain very similar.

6.2.1.4 Design Recommendations

Once an understanding of the pilots' actions had been collected using the SWARM technique and the PCM, it was possible to use this information to generate design concepts that would enable easier access to appropriate information. This aimed to reduce the likelihood of the suspected oil leak having a significant impact on flight operations. Table 6.5 lists a number of possible design recommendations that could be implemented on the flight deck that may reduce the requirement for, and number of, in-flight engine shutdowns as a result of abnormal engine system parameters. These recommendations are based upon providing the flight crew with reliable and accurate data relating to the on-going status of engine system parameters using the EMA.

As suggested in Table 6.5, an additional potential benefit relating to the diagnosis of an oil leak is supporting pilots in choosing an appropriate route diversion and an appropriate airfield for landing. Diversions are typically less of a problem for short-haul operators, operating in heavily populated and serviced locations, such as Europe and the United States, whereby maintenance bases are easily accessible. Yet, it is important to remember that choosing an appropriate, non-maintenance landing site is based upon multiple factors. These include environmental factors, for example local terrain and typography, local weather, runway length, type of approach and air traffic, aircraft status, including remaining fuel, weight of aircraft and type of aircraft, and commercial factors such as the location of maintenance, costs associated with displaced passengers and crew, availability of alternative accommodation and transport options for passengers. Thus, deciding where to divert is not always an easy and straightforward decision. Novel flight deck technology may be of value within this circumstance to assist pilots in determining the best diversion alternative. For instance, Table 6.5 suggests that appropriate diversion airports, be this due to aircraft type and weight or company preference, could be automatically highlighted on a new display and updated in real time based upon specific contextual requirements and location, reducing pilots' workload when dealing with an abnormal flight scenario.

6.3 CONCLUSIONS

This chapter has provided an overview of two approaches that can be used to generate design requirements alongside two example case studies.

The first approach seeks to identify the aspects of current technology that can be improved through modelling the tasks and subtasks involved in achieving specific case studies and then reviewing the opportunity for errors and failures. The case study example of the FMS showed how a redesign of the system and its interface could reduce the number of errors that can be experienced within the current technology. These modelling techniques are highly useful within the early stage of the design process to break down the comprising tasks and identify new ways of functioning.

The second approach suggests how the requirements for novel technologies can be generated through engaging with the desired end-use population. Interviews with

TABLE 6.5

Design Recommendations for the Flight Deck Instrumentation

Current practice	Possible deficiencies in performance	Design recommendations
Pilots deliberately check system displays and begin trending oil	Subtle leaks difficult to detect using current displays	Provide a graphical representation of oil trend data over time. Enable subtle abnormalities to be highlighted earlier
Determine how much time is left before engine reaches starvation by trending oil	Requires potentially complex estimations to be carried out by pilots in addition to standard flight tasks	Automatically present estimated time left to reach starvation based on flight parameters
Determine most appropriate strategy based on available information	Pilots may select sub-optimal strategy as they lack all available information or find it difficult to access information due to increase in workload as a consequence of the leak	Provide indication of preferred action but present various options
Follow relevant checklist from Quick Reference Handbook	Time lag associated with finding relevant checklist within Quick Reference Handbook	Automatically present relevant checklist for chosen action
Review information and check that any action has yielded the desired response	Difficult to maintain a detailed and accurate understanding of oil trend levels, especially when considering standard oil dynamics	Enable flight crew to review historic data and provide them with a projection of future engine state
	Pilots may neglect monitoring of healthy engine to focus on defective engine	Present data on healthy engine so that regular monitoring can be performed
Check diversion airports on the flight plan	Cost–benefit analysis of alternative airports based on commercial pressures may result in suboptimal choice	Automatically display most appropriate diversion airports based on aircraft type and other situational requirements

pilots identified how they currently assess aircraft engine oil levels and how a future system may be able to enhance the information that they have available to them. Using an engine oil leak as an exemplar scenario, the interviews sought to provide insight into the current procedures pilots must undertake, their thought processes and the sources of information that pilots access when dealing with such a scenario. Not only did the SWARM interview prompts enable the production of a perceptual cycle model for this process but they also enabled researchers to generate insight into how pilot decision-making can be better supported in such events. In addition to this, the same approach was used to determine how pilots may want to be assisted in the future and from this key design recommendations were generated. This represents a much more proactive approach to technology, and specifically cockpit display, design that engages end-users in the design process at a much earlier stage of the design lifecycle than is typical. Such engagement can lead to designs which are more suitable for implementation and installation within a cockpit environment and can more directly address end-user needs.

Once the requirements of the technology have been identified the next stage in the process is to begin designing the system. The following chapter will present the approach to generating initial designs using workshop techniques that are interactive and involve input from the users, manufacturers and Human Factors practitioners.

7 Design Generation

7.1 INTRODUCTION

This chapter outlines the application of methods to support the design and development of new flight deck technologies. Critically it presents how end-users, in this case airline pilots, can be heavily involved in the generation of design concepts. The 'Design with Intent' (DwI) method is presented to show how novel, yet functional, designs can be generated. The method is discussed in relation to engine warning systems within the cockpit, as they have previously been highlighted as being outdated and limited by current technological and architectural constraints. The benefits of applying user-led design approaches is discussed, as well as feasibility and logistical considerations. The second part of this chapter will compare the design ideas generated from the DwI method to the design recommendations generated by Human Factors practitioners using the Systemic Human Error Reduction and Prediction Approach (SHERPA), which seeks to develop designs through exposing errors in current systems (as has already been presented in Chapter 6). Through comparing insights from DwI workshops to SHERPA outputs, similarities and differences are revealed that highlight the utility of the two approaches as well as what can be gained from a combined approach.

7.2 GENERATING USER DESIGNS

Ideation is a creative process in which individuals are challenged to generate as many ideas as possible in order to address a specific problem statement. Importantly, ideation should be supported and completed in a judgement-free environment.

> *'It's not about coming up with the right idea, it's about generating the broadest range of possibilities'*
>
> Plattner, 2010, p.3

7.2.1 DESIGN ELICITATION TECHNIQUES – DESIGN WITH INTENT

Whilst there are a variety of methods available to facilitate the generation of design concepts, this chapter focuses on the Design with Intent toolkit (DwI) which aims to support the creative development of both physical items (i.e., interfaces) and behaviour. It was originally developed by Lockton et al (2009, 2010), who recognised that designers lack guidance on choosing appropriate design techniques to influence end-user interaction. For more information on this methodology see Chapter 3, Section 3.5.

7.2.1.1 DwI Requirements

Table 7.1 highlights a minimum set of requirements, in terms of input, equipment, personnel, time and expected outputs. From a practical viewpoint, it is important to ensure that a broad range of users from the commercial pilot cohort are included (i.e., it is important to consider aircraft type, culture, gender, experience level, etc. to assess how far design requirements may differ between different subsets of the population). Logistical challenges (e.g., location of workshops, costs associated, duration of workshops) must also be carefully considered.

7.2.1.2 Domains of Application

DwI has been predominantly applied within sustainability contexts (e.g. Lockton et al, 2013). However, Lockton (2017) advocates its use as a tool for interaction designers more widely and this chapter demonstrates its successful application to the design of novel flight-deck technologies.

7.2.2 Generation of a Novel Engine Health Warning System

7.2.2.1 Case Study: Oil Starvation Alert

The case study that was presented in Chapter 6 (Section 6.2) is examined again to demonstrate the application of the DwI methodology in the development of a new technology to assist pilots in managing oil leak events. Please refer to Chapter 6 Section 2 for more information on the oil starvation case study.

7.2.2.2 Participants

Designing novel interface concepts for a hypothetical system can be particularly challenging. This is because there is a tendency within the aviation industry to rely on test pilots to act as subject matter experts (SMEs) in all scenarios. However, in order to fully understand the interaction between the pilot and any new technology on the flight deck, it is important that research is conducted with a representative pilot cohort. In this case, commercial airline pilots represent the end-user of new flight-deck technologies. Relying solely on test pilots may bring a considerable risk that the expected benefits of a new technology may not come to fruition in reality (Damodaran, 1996). This is because test pilots are unlikely to behave like an average commercial airline pilot.

A total of five commercial airline pilots were recruited to take part in this study (three female, two male). They were aged between 31 and 38 (M = 34.6, SD = 2.7) and had held their licences for approximately 9.7 years (SD = 0.45). They had logged an approximate 3000–5000 hours (M = 4140, SD = 864.9) during this time and had

TABLE 7.1
DwI Requirements

Input	Equipment	Personnel	Expected Outputs
Scenario developed from Concept of Operations (CONOPs)	• DwI cards • Pen/paper for drawing	• Human Factors practitioners (facilitators) • Commercial airline pilots (users)	• Novel design concepts • Discussion on individual DwI cards

experience in flying with multiple airline carriers over the course of their careers including:

- Easyjet
- Tui
- Scandinavian Airlines
- Denim Air
- Olympic Air
- Air Atlanta Icelandic
- KLM Royal Dutch Airlines

7.2.2.3 Procedure

Upon arrival, participants were invited to read the participant information sheet and sign a consent form. They were then provided with a brief introduction pertaining to the purpose of the design workshop and the activities that would be involved. Participants were presented with a hypothetical scenario relating to an engine oil leak that would be the focus of the discussion:

> *'During normal operational flight, you are alerted to a suspected oil leak following a warning signal on the flight deck. You are aware that this may be a spurious alarm so must check to see if...*
>
> *1. The warning signal is valid*
> *2. Determine the criticality of the leak (i.e. trend oil system parameters)*
> *3. Take appropriate action'*

For the purposes of discussion, participants were told to assume that the initial warning was valid and they were asked to think about what information they would require to respond appropriately to this event and how this may be presented to them. Participants were then invited to draw 'initial designs' for the interface of a new pilot decision aid, either independently or as a group.

Upon completion of this activity, 42 DwI cards were presented to the group in a singular manner to encourage further discussion. This represents the 'prescriptive mode' of application that enables facilitators to choose the most appropriate cards to their problem (Lockhart et al, 2010). Down-selection was made by two Human Factors experts who assessed the cards for relevance to the scenario in question. For instance, cards relating to gamification were not included as these were not deemed relevant to the scenario under investigation. In addition, cards relating to the manipulation of human emotion were also removed as they were also deemed to be inappropriate in this context. At least one card from each lens was included.

Throughout the discussion, participants were invited to modify any drawings, create new ones, and note down anything that came to mind. Once all of the cards had been discussed, participants were invited to draw a 'final design' based upon the discussions that had taken place. It was up to the discretion of participants as to whether this final task be completed independently or as part of a group.

7.2.2.4 Outputs

Four unique design concepts were generated as part of this study. All stated the requirement for pilots to be provided with an indication of time and oil trend (i.e., time left before absolute minimum/maximum levels reached). Alternative perspectives on how this may be achieved were developed across the workshops (see Table 7.2).

TABLE 7.2
Alternative Design Options for the Presentation of Time Remaining Before Absolute Minimum/Maximum Levels Reached (Banks et al, 2019)

Idea		Notes
# 1		Lines on gauge can be used to show change in system parameter
# 2		Graph representation that incorporates a prediction of downward trend based on current system parameters in real time
# 3		Time displayed via some form of 'count down timer'

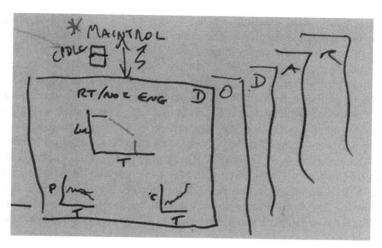

FIGURE 7.1 Proposed menu structure aligning to existing decision-making tools used by some commercial airline pilots.

There were also some notable differences, all with individual merit. For instance, some participants stated that it would be helpful to be automatically presented with the most relevant checklists and/or pages within the Quick Reference Handbook. Others suggested that the interface should align to current training processes surrounding decision-making tools such as DODAR (Diagnose, Options, Decide, Assign tasks, Review; Walters, 2002). One example of how this might be presented visually is presented in Figure 7.1.

Due to commercial sensitivities, it is not possible to share the final visual representations that were developed. However, a brief description of each concept is provided below to demonstrate how different groups approached the brief.

7.2.2.4.1 Concept 1

Final concept did not deviate too far from the traditional engine displays currently in existence. One reason for this may have been due to the cohort of pilots themselves – all predominantly flying short haul over Europe. In the event of abnormal engine oil parameters during such flights, their local knowledge would enable them to quickly identify suitable diversion airports. However, they did propose that a secondary page containing numeric trend data over time may be useful. This could be selected at the pilot's discretion. Typically, these are the sort of data that the pilots would be noting down so that they can calculate the time they have available. If this could be automatically presented, it may help reduce pilot workload in dealing with such an event.

7.2.2.4.2 Concept 2

As per Figure 7.1, the final concept utilised a popular decision-making framework to structure the response to such an event. DODAR (Diagnose, Options, Decide, Assign tasks, Review; Walters, 2002) is a form of memory aid that is designed to assist pilots in dealing with uncertain situations. On the 'Diagnose' page, a graphical depiction of trend may be useful to show oil system parameters over time. On the 'Options' page, a number of alternative responses may be proposed to the crew. The most preferable option could be presented in a larger font but pilots should still be free to choose from a number of alternatives. As an addition, it was proposed that some form of simulation to predict how such an action may impact upon oil system parameters may be presented. Further, it was recognised that option generation should take into consideration other key factors (e.g. location of maintenance, hospitals, and aerodrome requirements). One of these options could then be selected on screen (to satisfy the 'Decide' process of DODAR). The 'Assign tasks' could provide a digitised checklist for pilots to perform. Finally, the 'Review' page could provide a summary of the results following an action being implemented. Again, graphical depictions may be useful here but it may also be useful to include alternative options (e.g., 'divert') for if the situation changes. Each page could be accessed via interaction with touchscreen tabs.

7.2.2.4.3   Concept 3

This design consisted of two pages. Pilots suggested that traditional analogue displays should be entirely replaced with new, modern displays (e.g., bar charts). Page 2 would consist of a schematic diagram of the engine oil system, identifying exactly where in the system the apparent issue is located. One of the main ideas put forward relates to an information page (or 'i-page') that could provide more DODAR-related information via a pop up screen. This could be accessed via a single button press of an 'i' icon located in the top right-hand corner and removed by pressing an 'x' in the top right-hand corner of the pop up. Specifically, data to be incorporated on this page might include:

- Weather
- Predicted time remaining before starvation occurs
- Aerodrome requirements
- Runway distance requirements
- Weight requirements
- Limitations to landing (e.g., category)

Further, the 'i page' could also be used to provide a list of appropriate alternative airports based upon current level of criticality. Participants identified it may be helpful to identify a company preference but enable pilots to make the final decision. Any decision should be automatically fed back to the company via a link to the flight plan.

7.2.2.4.4   Concept 4

This concept was centred on the idea that information should be staggered (e.g., via a sequential tree-like diagram). An initial warning trigger, or 'attention getter', would be followed by pilot query. It was an acknowledgement that any warning messages should be presented only to the Pilot Monitoring (PM) in the first instance. This is because the primary concern of the Pilot Flying (PF) is to maintain control of the

TABLE 7.3
**Design Recommendations Yielded from Different Design Lenses
(Banks et al, 2019)**

Lens	Summary of design recommendations
Architectural	1. Standardised menu navigation is important to anchor in critical situations 2. Include automatic communication with the maintenance teams 3. Automatic pop up of relevant Quick Reference Handbook information 4. Enable movement of display (e.g., via touchscreen) to show others 5. Only present additional information when required
Error proofing	1. Maintain guarded switches for irreversible actions 2. Touchscreens should be used for confirmation actions rather than manual action buttons 3. Maintain aviate, navigate, communicate tasks 4. System should automatically consider variables (e.g., weather, runway length) when presenting alternative routes
Interaction	1. Only present oil information when it becomes critical to the continuity of the flight 2. Give estimated time until oil starvation to change expectations and evoke action 3. Enable users to amend actions based on new information (e.g., action suggestions from the maintenance teams via improved communication tools) 4. Provide a simulation/prediction of what will happen if certain actions are taken
Ludic	1. Utilise DODAR to structure the menu rather than current checklist representations 2. Group relevant information to facilitate next action
Perceptual	1. Remain consistent and comply with traditional colour conventions 2. Enable a comparison to be made between both engines 3. Provide options and make the preferred one larger, although pilot should still have freedom to choose 4. Addition of a camera or a diagram representation of the oil system may assist in the 'diagnose' phase of DODAR
Cognitive	1. The system should not commit pilots to an action, only guide them 2. Use of charts to represent oil chart
Machiavellian	1. Automatically place relevant screens in the right place and allow these to be moved freely by the pilot (i.e., dark cockpit scenario) 2. Match design to training 3. Telling the pilot they have less oil than they do could change decisions and prevent oil starvation
Security	1. Provide different information/options based on where you are in the flight, or the terrain you are flying over – recognise that different decisions will need to be made depending on context 2. Provide a map display with information about the locations and order options for alternatives based on your current position and requirements

flight. Therefore, data presented to the PM should be focused on supporting them through the diagnosis phase, followed by the presentation of different options that are available to them (e.g., fly idle, divert, do nothing, etc.). It was proposed that some information could be greyed out, providing an indication of both progress through the menu and also where the menu was going. Unlike the other concepts, this idea was based on a reconfigurable workspace, presented on a single screen. It was therefore envisaged to take up the entire span of the flight deck. Pilots would have the ability to 'swipe' away and move information around the screen. 'Pop ups' with relevant information may be utilised, but only at relevant times during the scenario. Thus, there must be an option for pilots to remove such pop ups (e.g., via an 'X' in the top right-hand corner).

To summarise, a number of design recommendations for onward development of the Engine Health Warning System were generated based upon the outputs of the DwI workshops. Examples of such recommendations are presented in Table 7.3, along with an indication of which lenses prompted these recommendations.

7.3   APPROPRIATENESS OF USER-LED DESIGN APPROACHES

Engaging with end-users during the earliest phases of the design lifecycle has provided the opportunity to generate and capture accurate user requirements. Further, using the process of participatory design (Damodaran, 1996), it has been possible to deliver insight into the types of information that flight crews may actually want to be presented with during abnormal operating scenarios as well as insight into the ways in which a new Engine Health Warning System may be used.

The DwI toolkit has proven to be instrumental in bridging the gap between engineers, designers and actual end-users (i.e., line pilots) much earlier on in the design lifecycle than would otherwise occur. It has also provided a useful approach in generating novel ideas for future flight deck technologies.

Of course, more research would be needed to assess their utility within their context of use. At best, DwI provides the initial starting point in which Human–Machine Interface wireframes can be developed. Subsequent design iteration would be expected following further end-user feedback and evaluation activities.

7.4   REVIEWING USER DESIGNS AGAINST DESIGN REQUIREMENTS

The value of the DwI method in enabling novel, user-centred designs is evident. Yet, the 'out-of-the-box thinking' that the method facilitates needs to be understood in relation to the complexities of the system within which the design is to be integrated such that it does not lead to other issues or negative emergent behaviours. In order to assess the designs that were generated with the DwI method, they were compared to a Human Error Identification (HEI) method.

HEI methods are particularly valuable early within the design cycle to determine potential design-induced errors that may occur within human–machine interaction (Baber & Stanton, 1994). The findings of such analysis can then be used to provide measures with which to remedy the errors and inform design (see Chapter 6, Section 2 for more information). The System Human Error Reduction and Prediction Approach (SHERPA; Embrey 1986) is an HEI method that aims to analyse system performance and identify errors induced by human operators and/or the design of a system. Hierarchical Task Analysis is combined with the SHERPA taxonomy of external error modes (Embrey 1986). This allows for a task analysis of the users' interaction with a system and the identification of potential error failures within each task. Application of this method has already been presented in Chapter 6, with the development of design requirements in relation to the re-design of a Flight Management System (FMS). Here, the SHERPA is applied again, this time to assess engine condition monitoring in the event of an oil leak. It is applied to identify potential errors in the current system and also to generate recommendations for remedial measures to overcome the identified errors.

In contrast to the DwI method, the remedial measures generated by the SHERPA will relate to alterations to the current way of conducting tasks, rather than striving for radical and novel techniques. The disparities in these two approaches are not yet clear. Therefore, the outputs of a SHERPA are compared to the outputs of the DwI method, which are less constrained by previous systems. Reviewing the SHERPA and DwI outputs in parallel will identify how effective they are individually, noting similarities and differences as well as determining what can be gained by reviewing findings across methods.

7.4.1 SHERPA METHOD

SHERPA was applied to determine the possibility for error occurring during current practice in responding to a suspected engine oil leak in the aircraft and propose new design ideas that could prevent these errors. Interviews were conducted with current airline pilots to inform of the development of an HTA and identify the tasks involved in both detecting and managing the oil leak. Once the tasks required to manage the oil leak were understood, the possibility for error to occur within each of these tasks was assessed.

7.4.1.1 Participants

Six pilots (two female, four male) with an Airline Transport Pilot Licence (ATPL) or Commercial Pilot Licence (CPL) for fixed-wing aircraft were interviewed. This was the point at which data saturation was reached (Grady, 1998; Saunders et al, 2018). It was therefore deemed the cut-off for the number of participants required for the analysis. Participants ranged in age from 26–35 years ($M = 30.17$, $S.D. = 3.02$). Participants had an average of 3692 hours of flight experience (range = 2900–4500, $S.D. = 635$) and 8.08 years of experience since obtaining their pilot's licence (range = 5.5–10 years, $S.D. = 1.74$). Each pilot was reimbursed for their time spent conducting the study and any travel expenses incurred. The interviews were run in accordance with the research institutes Ethical and Research Governance Office policies (ERGO ID: 40619).

7.4.1.2 Procedure

Pilots were interviewed individually to obtain information relating to their response to a suspected oil leak in a current aircraft system. This utilised the Schema Action World Research Methodology (SWARM; Plant & Stanton, 2016), which was developed to obtain information from pilots surrounding their decision-making processes and can be applied to understand what actions are available to pilots. See Chapter 3, Section 3.4.1 for more information. Interviews were audio recorded and the resulting transcripts were used to inform the development of the HTA. The main goal that was used as the starting point for the HTA was to 'manage a suspected oil leak'. A total of 78 tasks were identified in the HTA. The bottom-level tasks from the HTA were then reviewed in the SHERPA.

The SHERPA error taxonomy was used to determine the possible errors that occur within each of these low-level tasks in the system and therefore manifest into errors within the system, more information on the taxonomy can be found in Chapter 3 (Table 3.1). For each of the potential errors identified, remedial measures were proposed. A current commercial airline pilot with over 10 years of flight experience then reviewed the SHERPA for the errors identified and the remedial measures suggested, in order to determine if they were viable and appropriate.

7.4.1.3 SHERPA Results

The SHERPA analysis identified a total of 108 potential errors when responding to a suspected engine oil leak, the most frequent error type being action errors ($n = 36$) that related to omitted actions or conducting the wrong actions. The most frequent specific error type, however, was obtaining wrong information ($n = 19$), which was a retrieval failure. The focus of this work is to compare the SHERPA and DwI outputs so the individual outputs of the SHERPA are not presented here, although further detail can be found in Parnell et al (2021a).

7.4.1.4 Mapping DwI Workshop Responses to SHERPA Errors and Remedial Measures

The SHERPA generated a complete list of possible errors that could occur when pilots are managing a suspected oil leak in the current system, as well as a complementary list of remedial measures through which design could address these errors. Conversely, the DwI workshops generated a number of design concepts from potential end-users. The outputs from these methods were compared to determine whether the remedial measures suggested in each were related, or conversely presented opposing ideas. It should be noted that it was not the intention of the DwI workshops to generate a list of currently possible errors, so this was not the focus of comparison. Rather, the focus of the comparison was the extent to which the design concepts generated by end-users addressed errors identified within the current system.

A table was constructed listing the errors identified using SHERPA, alongside their corresponding remedial measures. The generated design concepts from the DwI workshops, presented in the second section of this chapter, were then reviewed against this list. The review determined if the design concepts matched the remedial measures developed using SHERPA, conflicted with the remedial measures developed using

TABLE 7.4

Requirements of the SHERPA and DwI Workshops

Method	Input	Equipment	Users	Outputs
SHERPA	Pilot interviews HTA	Pen/paper (or computer design program) for the SHERPA	3× Human Factors researchers 1× Pilot SME	Predicted errors in the current system Remedial measures to overcome current system predicted errors
DwI workshops	Oil leak scenario	DwI cards Pen/paper for drawing	2× Human Factors researchers 5× Pilot SMEs	Novel design concepts Discussion on each DwI card relating to design requirements

SHERPA or were novel ideas that had not been identified using SHERPA. Table 7.4 presents the inputs, equipment, users required for the SHERPA analysis and DwI workshops as well as the expected outputs. This highlights that while the SHERPA utilises input from expert users to inform the development of the HTA, independent researchers predominantly drive the analysis. DwI, in contrast, is developed and driven by the insight of potential end-users.

Across the design concepts there were several pertinent ideas that consistently emerged, however there were also some conflicting ideas. The level of detail and insight that could be gleaned from involving pilots in the design process was evident, as well as their generation of solutions to the problems they identified from their perspective. An example excerpt from the table is presented in Table 7.5.

Using the complete table contrasting the DwI recommendations to the SHERPA errors and recommendations it was possible to identify which ideas from the DwI were a match with the SHERPA, conflicted with the SHERPA or were new ideas outside of the SHERPA. The frequencies of each of these discrete occurrences were calculated and are presented in Table 7.6. The ∞represents all other ideas that neither method generated, the potential of which cannot be accounted for in this analysis.

Table 7.6 gives an indication that there was some overlap in the design recommendations that the two methods proposed and thus demonstrates that using SME participants to generate ideas independently of HEI methods can validate the outputs. Yet, there were also recommendations that each method proposed that were not supported by the other. A key area of concern here is the nine recommendations made using SHERPA that conflict with the SME recommendations made in the DwI design workshops. This suggests that SHERPA alone may not be an ideal method of capturing design measures that are practical and useful to the user of the system. That is to say we cannot rely solely on analyses conducted without the involvement

TABLE 7.5
Extract from the Table Contrasting the Dwi Recommendations to the SHERPA Errors and Recommendations

SHERPA Error	SHERPA remedial measures	Match	Dwl Conflict	Dwl New ideas
Fail to compare oil level/temperature/ pressure to correct limits	Bring up current oil level/ temperature/pressure (bar display) with historic and predicted oil levels (graph display)	"i button" only appears when needed and then it gives information on the next phase of flight that is required. Location of the oil level on the dial needs to be easy to read	Pilots need freedom to look at what they want	If something more important happens, it needs to be able to be overridden (task lock in/out – pop up checklists with actions). Different pages of information and not presenting too much information on one page to prevent clutter
Fail to adjust calculation of trend for remaining flight	Automatically trend current and predicted oil parameters in response to updated flight parameters (e.g. change in throttle)	Option of reviewing the predicted trends across different time periods, e.g. 10 mins or 30 mins. This would give different rates that may be required for different types of flights. Real-time feedback could be given based on their action	Too complicated to simulate and understand every possible option. Keep it simple	Count down timer on how much time you have left under oil starvation. Rate number could be added to the dial on primary screen

TABLE 7.6

Frequency of Ideas Generated in the DwI that were Present or Absent in the SHERPA and Vice Versa

	DwI	
SHERPA	**Match**	**Conflict**
Match	**29**	**9**
	Proposed in both the SHERPA and DwI	Proposed in the SHERPA but not the DwI
Conflict	**96**	∞
	Proposed in the DwI but not the SHERPA	Not proposed in DwI or SHERPA

of representative end-users. It is also interesting to note the 97 new design ideas that could resolve the errors identified using SHERPA were generated from DwI discussions but were not identified using SHERPA. This demonstrates the rich data generated by the DwI workshops, and further supports the value of end-user input into design generation. To illustrate the impact of the differences and similarities across the two methods, a case example is presented.

7.4.1.4.1 Case Study of SHERPA Error: Fail to Adjust Calculation of Oil Temperature/Pressure Leak Trend for the Remainder of Flight

SHERPA remedial measure: Automatically trend current and predicted oil parameters in response to updated flight parameters.

DwI similarities

Pilots identified the need to update the predicted oil pressure/temperature levels in line with the flight parameters. It was suggested that the option of having access to information related to the trend of the oil over different time periods and in relation to different parameters depending on the status of the flight would be beneficial. They suggested that real-time feedback on the oil leak status could be given in response to the actions that they carried out in their attempt to manage the situation, for example reducing the throttle and powering down the engine.

DwI differences

Divergent ideas also emerged, with some pilots suggesting that the option to simulate every possible option and possibility would be too complicated to process and understand in a scenario such as this. Pilots cautioned the proposition of overly complex information and the presentation of the multiple possible actions they could take, favouring instead easy-to-understand real-time feedback on actions they had taken.

DwI new ideas

Pilots promoted the need for information regarding the 'time until oil starvation' and were therefore keen to have this information presented to them in a clear and easily accessible manner. While access to detailed information on the oil leak trend on secondary displays was suggested, some pilots also suggested bringing information on the rate of change or the 'time until oil starvation' to the primary display when there was a suspected oil leak, as described in this scenario. This could involve a count-down timer on the main display or an indicator of the rate of oil leak on the primary oil level display. Noting that if there was no oil leak this information would not be needed, and therefore should not appear, on the primary display.

7.4.2 DESIGN INSIGHTS FROM HUMAN FACTORS PRACTITIONERS VS. DOMAIN EXPERTS

Despite the clear accessibility and advantages gained using SHERPA, it was also evident that end-users, i.e. airline pilots, can propose ideas that counter those made by Human Factors researchers using this approach. In addition, it was also evident that end-users are capable of generating a wealth of other ideas that may greatly assist the design process. These novel ideas can be invaluable to the design process but are frequently ignored when the end-user of the system is not represented within the overall design process (Kujala, 2003). Through mapping the reports that the pilots gave within the DwI workshops to the errors and remedial measures that the SHERPA approach identified, it was clear that SHERPA was effective in generating usable remedial measures. However, it is also clear that end-users would not have approved of some of the developed remedial measures. Notably, it was evident that pilots have an expert's insight into how a future system would be integrated alongside all the other features and tasks that must be completed, of which SHERPA analysts would not be fully aware of.

The disparity between those with domain knowledge and those with Human Factors knowledge in conducting error analysis is suggested by Stanton and Baber (2002). The judgement of the analyst plays a large role in the output of HEI methods application, yet Stanton and Baber (2002) identified that novice users of a system could apply such methods to an acceptable standard with ease. This was deemed to be due to the structured nature of the SHERPA and Task Analysis for Error Identification (TAFEI) that enabled a structure for judgements to be made, without constraining them (Stanton & Baber, 2002). The findings from the current work demonstrate that while the Human Factors researchers were able to adequately predict possible errors within the system and determine remedial measures that mitigated these errors, final concept generation is much improved with the addition of end-user input. The insights offered by potential end-users added rich detail in what would increase engagement with a future system and its integration within the flight deck as well as guiding design steps. This is an important consideration that practitioners without user expertise should be aware of when applying Human Factors methods within a focused design approach. Overall, it is apparent that a combination of insight from both Human Factors practitioners and end-users would add the most value to the development of novel systems.

7.5 SUMMARY

This chapter has shown how input from end-users within design workshops can add valuable insights when applying SHERPA and provide recommendations for error prevention. The insights that Human Factors researchers gain from the application of these methods, even when they are validated by an SME, omit substantial detail regarding the wider functionality of the system and the implications it may have within this. Through the application of the DwI method, greater insights into the preferences of the pilot user of the system have been obtained that have highlighted where recommendations made through the SHERPA may be effective and conversely disruptive. For the generation of usable and error-resistant interfaces it is important the Human Factors methodologies are true to the preferences of the user and the functionality of the wider system as a whole.

The next chapter will focus on the next stage of the design process: modelling the designs. This involves reviewing the designs within the wider environment within which they will be used, the interactions that they will require and how they will be used by the intended user population.

8 Design Modelling

8.1 INTRODUCTION

In the previous chapter a number of designs were generated using user-centred design principals. This chapter will present ways in which such designs can be reviewed to understand if they will meet requirements. It will detail the importance of modelling new designs at the early stages of the design lifecycle, after conception but pre-development, to determine if they can provide enhanced performance and minimise opportunity for error.

This chapter will discuss two key areas of modelling which are useful to the design process; engineering integration modelling and user behaviour modelling. Modelling engineering integration involves modelling the system within which the design will function to understand the interaction with other system components and the functionalities that will be performed through these interactions. This allows an insight into the new ways that systems may need to perform or adapt with the addition of new designs and technologies.

The second type of modelling that will be discussed is user modelling. This involves modelling user behaviour and will concentrate on the application of theoretical models to capture and understand decision-making.

8.2 ENGINEERING INTEGRATION MODELLING

There are various approaches that can be used to understand the impact of new technology on traditional ways of working (see Stanton et al, 2013 for a comprehensive summary of each approach). However, we have found that Operator Event Sequence Diagrams (OESDs) are particularly well received within the systems engineering community. OESDs are widely used here to explore the relationships between individual subsystem components (Banks & Stanton, 2018). They originate from the weapons industry, whereby explicit human–machine interaction needed to be defined (Brooks, 1960). However, they have since been more broadly applied to a wide range of domains including flight decks, air traffic control (Harris et al, 2015; Huddlestone et al, 2017; Sorensen et al, 2011) and automated driving (Banks & Stanton, 2018).

OESDs are essentially a process chart, detailing the sequence of tasks that must be completed as well as the interaction between humans and technological artefacts

DOI: 10.1201/9781003384465-8

TABLE 8.1
OESD Key

Symbol	Description
	Process
	Display/ information
	Data/ Received information
	Manual Operation
→	Connector
	Terminator
	Decision

(Stanton et al, 2013). They can be used to facilitate comparisons between different human/machine configurations (i.e., identify tasks and functions that are no longer required) as well as identifying new automation requirements (Harris et al, 2015). They use a set of standardised geometric shapes to portray meaning (see Table 8.1). Readers are encouraged to familiarise themselves with the meaning of these symbols to make sense of the OESDs within this chapter.

The process of developing OESD is discussed further by Stanton et al (2013), and is summarised in Figure 8.1.

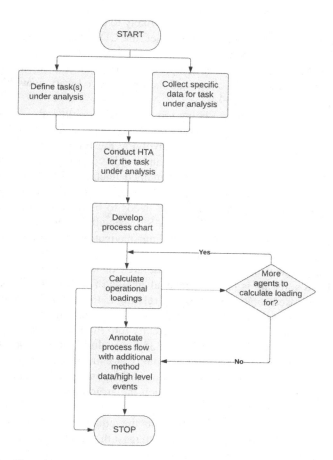

FIGURE 8.1 Flowchart outlining process of OESD development (adapted from Stanton et al, 2013).

8.2.1 Case Study: Application of OESD to Next-generation Flight Management Systems

Flight Management Systems (FMS) are one of the most sophisticated technologies on-board modern-day aircraft (Avery, 2011). The FMS determines the position of the aircraft and aids in adherence to a flight plan. The main input modality into the FMS is via a Control Display Unit (CDU) (Singer & Dekker, 2001). More information on the FMS and information relating to modelling of FMS capabilities can be found in Chapter 6 (Section 6.1.2).

Whilst the design of the CDU is typical to all manufacturers (i.e., it includes a screen, line select keys, alphanumeric keys, function keys, and warning lights), there are some differences with regards to the position of letter keys and use of function keys across manufacturers. This brings with it the risk of 'negative transfer' when pilots transition between systems (e.g., more likely to press the wrong button; Singer & Dekker, 2001; Stanton et al, 2009). Further, the CDU has multiple layers, which creates a 'keyhole effect' (Woods & Watts, 1997) whereby users can get lost within the system, as they can only see a small portion of relevant information at any time. Thus, pilots must memorise functions and their respective locations within the menu structure (Marrenbach et al., 1998).

The FMS is often criticised for being too complex and vulnerable to human error (Courteney, 1998; International Air Transport Association, 2015). In 1992, Eldredge et al reported that the most likely errors involving the FMS centred on flight crew misinterpretation of data, keyboard errors and mode control panel selection errors. Given that the design of the FMS has remained unchanged for decades it seems likely that the same errors continue to occur (International Air Transport Association, 2015). Billings (1991) suggested that more needs to be done to improve the inter-action between the flight crew and the FMS, especially given that it is integrated into virtually every flight function, including: navigation, performance, guidance, display management and data management. One possible improvement could be to utilise technological advancements in the area of increased connectivity. Next-generation FMS software is likely to take advantage of such innovations. In August 2018, Boeing released some initial findings relating to their new 'RouteSync' service (Boeing, 2018). 'RouteSync' enables the automatic uplink of flight plans, weather information and other performance data directly into the FMS. Boeing (2018) claim that 'RouteSync' saved between 200 and 335 hours during a 6-week test period as the requirement for pilots to perform manual data entry tasks is reduced consider-ably. Similarly, GE Aviation (2018) introduced the concept of a 'Connected FMS' (CFMS) which provides a platform in which tablet applications can interface dir-ectly with the aircraft FMS. One specific capability of the CFMS is to automatically update flight plans to the FMS, again removing the requirement for pilots to manually input data. These technological innovations suggest that a pilot's interaction with the FMS is therefore set to change as increasing levels of automation are introduced into the system. However, it is unclear how the new interface between the pilot and the connected FMS will appear and behave.

Lorenzo-del-Castillo and Couture (2016) speculate future cockpit design to utilise more interactive, tactile inputs (i.e., incorporating tactile surfaces, gesture interaction, voice recognition and position detection). This is consistent with the way in which we interact with a whole host of everyday products including smartphones, home appliances and ticket-selling points (that have moved away from physical control inputs to touch-enabled surfaces). With this in mind, it seems reasonable to assume touchscreen technology will become a more prominent feature within the flight deck. There has already been a vast amount of research conducted that assesses its utility within the flight deck environment (e.g., Coutts et al, 2019; Dodd, 2014; Harris, 2011; Stanton et al, 2013).

Nowadays, the majority of airline pilots use an Electronic Flight Bag (EFB). Portable EFBs come in multiple forms (i.e., smartphone, tablet, laptop) and are used by pilots to perform a variety of functions that have previously relied upon paper-based documents (Winter et al, 2018). Currently we see great variability in the level of information integration, with more advanced EFBs able to display an aircraft's position on navigational charts, show real-time weather information and be used to perform complex flight-planning tasks. Conventionally, some EFBs (referred to as installed EFBs) are built into the aircraft's architecture. However, it may be possible that, in the future, some pre-flight tasks relating to route planning could be conducted away from the aircraft (e.g., services provided by RocketRoute, 2018). Skaves (2011) identified additional functionality for portable EFB including, but not limited to, uploading flight planning information directly into the aircraft FMS. Currently EFB can be used in the aircraft cockpit but they are not recognised as certified equipment of the aircraft (Turiak et al, 2014). Skaves (2011) has argued however, that operators have sought additional EFB capability for some time as well as expressing a desire to also expand the scope of operational use. Connected EFB therefore offers a pertinent avenue in which the operational capabilities of EFB can be extended and accepted by end-users (i.e., pilots).

8.2.1.1 Comparing 'Present' with 'Future' Methods of Interacting with Flight Management Systems

Two OESDs were developed to portray FMS system initialisation (i.e., pre-flight route planning activities) based on modern day (i.e., 'present') and future operations. Four Human Factors practitioners and a commercial airline pilot with 10 years flight experience who acted as a subject matter expert (SME) were involved in their development. The commercial airline pilot had over 5000 recorded flight hours on Boeing 757, 767 and 787 aircraft and so it was these systems the 'present' OESD was based upon. The use of an SME to confirm the procedures and processes is a common practice in Human Factors research (e.g., McLean et al., 2019).

The first step in constructing the OESD was to identify all possible system agents that are involved during the system initialisation process (see Table 8.2). Whilst the predominant method of data entry in recent years has been via Datalink or pre-programmed company routes, it is still important to focus on manual data entry because it is recognised that in situations whereby the plane is outside the range of Aircraft Communications Addressing and Reporting System (ACARS), the flight crew must still enter data manually. Given that human operators are often seen as the last line of defence (Dekker, 2014), it is important that the manual task of data entry is considered as it is not an entirely redundant input strategy.

Online video tutorials relating to the operation of the FMS were used to facilitate the construction of an OESD for current practice. Commercial videos for future systems were used to inspire OESD development for future systems which were accessed through online forums such as YouTube (e.g., Boeing, 2018; GE Aviation, 2018; Mentour Pilot, 2015, 2017).

Figure 8.2 provides an excerpt of the OESD created for current means of interacting with the FMS. It identifies the tasks and operations that are carried out by each corresponding agent and demonstrates how these may be linked, or interact, with other agents within the system network. It should be viewed as a prototypical representation of how a pilot may operate the system as there are a number of ways in which system initialisation can be achieved (i.e. there is no standardised way of

TABLE 8.2
Actors Involved in System Initialisation Activities

Actor	Responsibilities
Pilot 1	Inputting, checking and cross-checking data within the FMS
Pilot 2	Checking and cross-checking data within the FMS
Electronic Flight Bag (EFB) 1 & 2	Contains all relevant data relating to route planning activities including information on the flight plan
Control Display Unit (CDU)	Enable pilots to interface with the FMS using a mix of soft and hard keys
Flight Management System (FMS)	Hold all data relating to navigation and performance – the 'heart of all automation in the modern commercial aircraft' (Harris, 2016, p. 238)
Navigation Display (ND)	Present track and heading against a simplified map
Global Positioning System (GPS)	Provide accurate data relating to location
Inertial Reference System (IRS)	Sense and compute linear accelerations and angular turning rates. Data are used for navigational computations
Checklist (C)	Provide guidance on tasks to be completed
Dispatcher	Generates the flight plan

interacting with the FMS). It is therefore important to acknowledge that minor deviation from this pathway of interaction may occur (although this only relates to the order in which data are entered, not the processes involved). Overall, a minimum of 142 operations were identified using the OESD approach that must be completed in order to satisfy the system initialisation process.

This is in contrast to a minimum of 80 operations when using alternative methods of system initialisation (an excerpt of the OESD for 'future practice' is presented in Figure 8.3). This suggests a significant reduction in workload requirements with the introduction of new technology. Rather than manually inputting data into the system, Pilot 1 simply needs to cross-check the data available on screen once the synchronisation process has been completed (i.e., in other words, perform a consistency check). Once Pilot 2 has entered the flight deck and completed the initial pairing activity, they would need to complete a 'silent' check between the pre-loaded FMS and Portable EFB 2 to ensure consistency before confirming system initialisation with Pilot 1. It is also worth noting that prior to embarking the aircraft, both pilots would need to check that both of their portable EFB devices are presenting the same information. This process ensures consistency between EFB 1 and EFB 2, and between EFB 1 and the FMS, and finally consistency between the FMS and EFB 2 which provides a more robust strategy than current operational practice. Thus, the portable EFB could become like the FMS is for automation, the heart of all route planning activities on the open flight deck.

In order to further emphasise the differences between 'present' and 'future' systems, it is possible to calculate operational loading scores. These scores simply

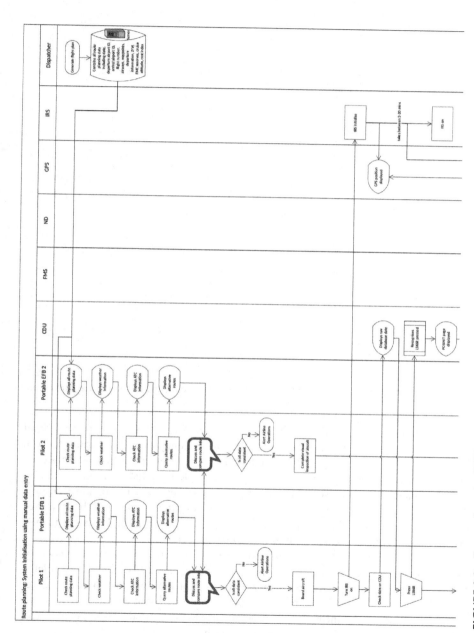

FIGURE 8.2 Excerpt from the OESD modelling 'current practice' in FMS system initialisation using manual data entry.

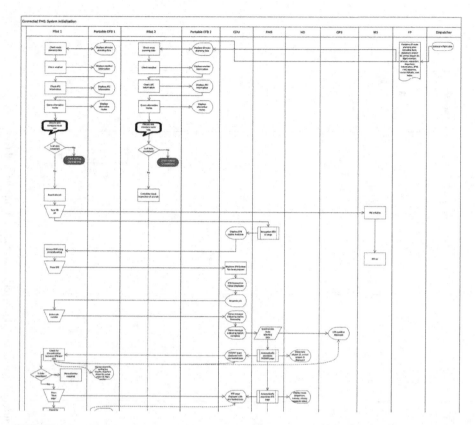

FIGURE 8.3   Excerpt from the OESD modelling future practice in FMS system initialisation.

provide a count of all operations within the task (e.g., process, decision, manual input etc.) (Stanton et al, 2013). Table 8.3 presents the frequency of each operational load. The greatest differences appear to be related to process and pilot task loading (i.e., reduction in manual inputs). However, it is interesting to note that alternative methods of FMS interaction may serve to slightly increase the decision-making burden. When we consider only the operational loadings associated with the flight crew (see Table 8.4), we can see that the task loading associated with Pilot 1 is significantly reduced when using a Connected EFB by approximately 63%.

There are, however, a greater number of decisions required by Pilot 1 as they conduct 'review'-based activities associated with a connected EFB, namely associated with ensuring consistency between devices. Instead of inputting data in the 'future' system, Pilot 1 must now make multiple decisions through-out the process to determine whether or not the information automatically loaded into the FMS is consistent with the data presented on EFB 1. This increases the potential for new types of error to emerge. Rather than action-based errors associated with inputting data (Marrenbach & Kraiss, 1998), there is greater potential for checking-based errors associated

TABLE 8.3
Operational Loading Scores for Present and Future Methods of Interacting with the FMS

Operation loading	Present	Future	Difference
Process	49	22	−27
Decision	2	6	+4
Sub-process	3	6	+3
Data	6	1	−5
Display	43	29	−14
Manual input	30	8	−22
Communication	4	4	0
Database	1	1	0
Document	1	0	−1
Start/end	3	3	0
Total	**142**	**80**	**−62**

TABLE 8.4
Comparison of Total Number of Operations Based on Pilot Role (E.G., Pilot 1 Versus Pilot 2) in 'Present' and 'Future' Systems

Operation loading	Pilot 1			Pilot 2		
	Present	Future	Difference	Present	Future	Difference
Process	30	10	−20	7	8	+1
Decision	1	5	+4	1	1	0
Sub-process	0	0	0	0	0	0
Data	0	0	0	0	0	0
Display	0	0	0	0	0	0
Manual input	30	6	−24	0	2	+2
Communication	2	2	0	2	2	0
Database	0	0	0	0	0	0
Document	0	0	0	0	0	0
Start/end	1	1	0	1	1	0
Total	**64**	**24**	**−40**	**11**	**14**	**+3**

with comparing two display screens. This task may place more burden on the pilot, increasing their workload and leaving them vulnerable to failure. Until the FMS is able to identify erroneous input (and subsequent output), the onus of responsibility remains firmly on the shoulders of the pilots (International Air Transport Association, 2015). Going forward, we may want to evaluate alternative functional allocations (e.g., designing software capable of 'flagging' inconsistencies between displays) to reduce the likelihood of checking-based errors. Notably, Pilot 2 sees a slight increase in task loading due to the requirements surrounding the linkage between EFB and FMS. This is unlikely to pose a significant cognitive burden, especially when we consider the context in which the work must be completed (i.e., pre-flight set up).

8.2.1.2   Engineering Integration Summary

The OESDs presented here predict how new flight deck technology may change the nature of FMS system initialisation from a manual data entry task to more of a checking and reviewing task. Checking and reviewing tasks in particular seem to be a prime candidate for automation support, especially if there is a move towards single crew operation. For instance, Harris et al (2015) suggested automation could be used to cross-check pilot actions against predetermined checklist requirements. This may also be applicable to pre-flight tasks involving connected technologies.

As connected EFBs have not yet been developed, these early conceptual models provide a foundation upon which the subsequent design and development can be based. In this case, future research would aim to validate these findings in a flight simulator that can represent the functions of a connected EFB.

More generally however, it is anticipated that developing OESDs early on in the design lifecycle will assist in the development of design and user requirements for future systems. However, they can be supplemented through the use of more user-based models. User-based models in contrast provide insight (and perhaps greater clarity) into the cognitive processing of system operators. Together, they are a powerful approach to understand task–user interaction. The next section will focus on decision-making and how different models of decision-making can be applied to the decisions that pilots need to make.

8.3   USER BEHAVIOUR MODELLING: HUMAN DECISION-MAKING

Human decision-making is a complex and well-established area of study, with multiple theoretical approaches. Of particular interest to this work is Naturalistic Decision-Making (NDM): the study of decision-making in association with the environment that the decision occurs and the decision-maker themselves (Gore et al, 2018; Klein, 2008). The field of NDM diverged from previous decision-making research at the time of its conception, which had assessed controlled, structured environments where the decision-maker was passive to the outcome (Klein, 2008). The inception of the field of NDM began with reviewing decision-making 'in the field' to understand the strategies behind natural decisions (Klein, 2008). Multiple different theories of NDM have since come to fruition (Lipshitz, 1993) and there is still much debate in the field over the best ways to conceptualise decision-making under naturalistic conditions (Lintern, 2010; Lipshitz, 1993; Naikar, 2010). However, three models have been selected for review to understand the variances in decision model theories and the implications that this can have for the design of technologies such as decision aids. These are the Recognition Primed Decision Model (RPDM; Klein, 1989), Decision Ladders (Rasmussen, 1983) and the Perceptual Cycle Model (PCM; Neisser, 1976). These three models were chosen as they can account for decision-making under critical real-world conditions and have been previously applied to the aviation domain (Asmayawati & Nixon, 2020; Banks et al, 2020a; Hu et al, 2018; Plant & Stanton, 2012, 2014; Stanton et al, 2010; Vidulich et al, 2010).

In this section we apply these decision models to interview data collected from airline pilots to understand how the models can comparatively inform the design of a decision aid that can assist the response to a dual-engine failure on take-off. This relates to a bird strike and engine fan-damage case study.

8.3.1 CASE STUDY: ENGINE BIRD STRIKE

An infamous aviation event is the Hudson River incident wherein US Airways Flight 1549 took off from New York and underwent a series of bird-strikes which caused significant damage to both engines and near loss of total thrust. The case has been made famous through the film 'Sully', capturing the actions of the Captain, Chesley Sullenberger. The response of the crew enabled the survivability of the incident when the pilots ditched the aircraft on the Hudson River. The National Transportation Safety Board (NTSB, 2010) report into the Hudson incident provided new safety recommendations to the FAA, the first of which stated the need for the development of 'technology capable of informing pilots about the continuing operational status of an engine' (NTSB, 2010, p. 124). Before such a system can be developed however, it is important that current flight deck processes are understood, which includes decision-making processes on behalf of the pilots. The integration of new functionality can then be assessed to ensure any new information is usable, effective and safe.

Here we will apply qualitative interview data from commercial airline pilots to each of the decision models to understand how they capture decision-making during a dual-engine failure on take-off. The models are then discussed in terms of how they can inform the design and development of a decision aid. We compare and discuss the contributions of each of the decision models and summarise key design recommendations.

8.3.1.1 Methodology: Pilot Decision-making Data Collection

To populate the decision models, interviews were conducted with eight commercial airline pilots with a range of experiences. The interview reports were then used as the data source for the application of the three decision-making models.

8.3.1.1.1 Participants

Eight commercial airline pilots were recruited to take part in this study (three female), aged 29–65 (M = 39.42, SD = 14.01). All participants were qualified fixed-wing ATPL or CPL pilots with an average 7085 hours flight experience (SD = 10231.72) having held their licences for an average of 13.60 years (SD = 14.28). The pilots had employment experience with 13 different airlines and at the time piloted a range of different aircraft including Airbus ($n = 3$), Embraer ($n = 3$) and Boeing ($n = 2$). The study was ethically approved, and participants were reimbursed for their time. Participants were recruited until a point of saturation in the data was achieved.

8.3.1.1.2 Equipment

The interviews were conducted virtually using the video conferencing platform Microsoft Teams. The participants were able to see the researchers while a PowerPoint presentation displayed the questions via the 'shared screen' tool. Microsoft Teams also allowed the interviews to be recorded in full for later analysis, subject to permission by the participant at the start of the interview.

8.3.1.1.3 Procedure

Two researchers conducted semi-structured interviews with one pilot at a time. Each lasted approximately one hour. During each interview participants were presented with the following engine failure scenario. This scenario features the same engine damage and altitude as the Hudson River incident (NTSB, 2010):

You are flying a twin-engine aircraft during its initial climb (~ 2800 feet). A flock of birds strike both engines. You must:

- *Determine the criticality of the situation (e.g., state of each engine)*
- *Take appropriate action*

The Critical Decision Method (CDM; Klein et al, 1989) was used to elicit knowledge in relation to situation awareness and decisions made within non-routine incidents. As detailed in the CDM procedure (Klein et al, 1989), participants were first asked to detail their initial thoughts and give an overview of the scenario in an open manner, with the option to ask for further clarifications and questions to the researchers.

8.3.1.1.4 Data Analysis

The recorded interviews were transcribed using the automated transcription function within the Microsoft Teams software. As the pilots gave relatively similar responses due to the training they receive, and because the interviews were conducted until the point of data saturation, the interviews could be aggregated in their application to the three different models. Application of the data to the three different models was done in accordance with the guidance in the literature for each method (Jenkins et al, 2010; Klein et al, 1989; Plant & Stanton, 2016). These models include the RPDM, Decision Ladders and the PCM. Validity assessments were made with experienced Human Factors practitioners and an airline pilot (with over 10 years' experience) reviewing and refining the models.

8.3.1.2 Recognition Primed Decision Model

The RPDM is a prominent decision model developed by Klein (1989). It has remained relevant over time and is still popular in its applications to decision-making in a variety of domains today (Hu et al, 2018; Neville et al, 2017; Yang et al, 2020). The RPDM proposes that decisions are made through the recognition of critical information and prior knowledge (Klein, 1989). If a situation can be matched to a previous event, prior experience will guide an individuals' interaction and any necessary decision-making. The RPDM states that decision-making is comprised of four key stages (Klein, 1989): Recognition, Situational Assessment, Mental Simulation and Evaluation. The Situation Assessment Record (SAR; Klein et al, 1989) is a way of mapping out decision-making points (see Table 8.5).

TABLE 8.5
Situational Assessment Record (SAR) of the Bird-Strike Event

SA – 1

Cues/knowledge	Hear, feel, see or smell birds hit the engine.
	Engine instrumentation, N1/N2, high vibration indication and/or high temperature
Expectations	Potential dual-engine damage, one engine may give some thrust, risk to flight safety, plane cannot maintain altitude with reduced thrust
Goals	Maintain the flight of the aircraft (manual or with autopilot) to assess available thrust
	Decision point 1: Activate autopilot (if not already engaged)
	Decision point 2: Cross confirm flight and engine metrics with co-pilot

SA – 2 (Elaboration)

Cues/knowledge	Engine indicators show engine parameters, ECAM message, available thrust, confirmation from co-pilot on engine damage, QRH/ checklists for engine shut down
Expectations	At least one engine is severely damaged, engine may be on fire, engine(s) may need to be shut down and isolated
Goals	Determine if the damaged engine(s) needs to be shut down, maintain enough altitude to reach a suitable landing position
	Decision point 3: Determine that engine damage is severe and at least one engine needs to be shut down and isolated

SA – 3 (Elaboration)

Cues/knowledge	ECAM message, engine indicators show engine parameters, QRH/ checklist for engine shut-down
Expectations	Shut down engine to isolate it and prevent further damage.
	Shut down the most damaged engine
Goals	Maintain enough thrust to return and land at departure airport
	Decision point 4: Return to departure airport

SA – 4 (Shift)

Cues/knowledge	Second engine may need to be shut down – ECAM and engine indicators
Expectations	Both engines are severely damaged and cannot generate thrust from the remaining engine
Goals	Shut down both engines; action emergency turn procedure; identify best available landing position
	Decision point 5 – Land the aircraft in best possible location with two damaged engines

SA – 5 (Elaboration)

Cues/knowledge	Pre-flight briefing identified the emergency turn procedure, navigation display and view outside identify available landing positions, QRH checklist for aircraft ditching
Expectations	Need to perform an emergency turn, may need to 'ditch' the aircraft
Goals	Perform emergency turn procedure; land aircraft safely

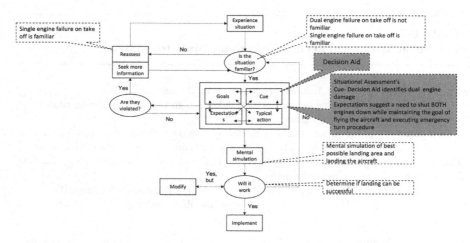

FIGURE 8.4 RPDM showing where the decision aid could provide assistance to the process of managing a dual-engine failure on take-off.

In the engine failure scenario the application of recognition strategies was limited due to the pilots' inexperience of dual engine failures. Instead, they relied on experience of single-engine failures on take-off and their knowledge of the Quick Reference Handbook (QRH) and checklists which outline the procedures for critical events. This assessment is shown in the first loop of the RPDM in Figure 8.4. Once they have established some recognition of a similar event, they move through to the process of assessing the situation.

Once the situation assessment is made, the RPDM in Figure 8.4 shows that mental simulation of the intended actions is generated by the pilot. Mental simulation involves determining if the intended landing site is feasible, given the perceived engine damage and environmental conditions. The RPDM states a serial process of considering options – however, there would be limited time for serial evaluation as the aircraft would lose altitude and begin to glide without thrust and power from an engine. There would be consideration of the available time, and if it is suitable to execute the emergency turn procedure, as identified in a pre-flight briefing. If not, additional serial evaluations would need to occur to identify a suitable alternative landing site.

In terms of understanding how a decision aid could be used to inform decisions, the RPDM highlights the importance of recognition in the alerts that pilots are given to quickly diagnose the situation. As recognition is the initial step in the RPDM process, recognition for the exact nature of the problem needs to come at the start of the decision-making process. The serial nature of option generation and selection further highlights the need to recognise the severity of the event from the time of the alert. Figure 8.4 also shows how a decision aid could be applied within the RPDM as a cue within the situational assessment. A decision aid that could provide the cue that both engines were damaged and clearly show the extent of the damage would enable the pilot to have accurate expectations at an early stage in the decision-making process.

This would limit multiple, serial cycles of the model, which is key in a time-critical event. Furthermore, a display that utilises principles that are familiar and recognisable as an engine failure are recommended.

The RPDM only requires 'good enough' (Simon, 1955) decisions to be made, rather than optimal ones (Klein & Calderwood, 1996). Yet, the aviation domain upholds stringent decision-making practices to ensure that its pilots act in a justifiable manner, in keeping with standards (Kaempf & Klein, 1994). For example, in the case of the Hudson incident, the FAA undertook a large-scale investigation into the actions of the pilots who then had to defend themselves in a court of judgement. Application of the RPDM must therefore consider the importance of the decision and its implications. In emergency situations, such as the one studied here, safety is the priority and other factors are often easily traded off. For example, the pilots said that the companies preferred choice of landing sites, or costs to the company would not factor into their decision-making due to the critical nature of this situation. The RPDM therefore captures how the assessment and goals of the pilots prioritise the passengers' safety. If further detail in the decision-making is required, then the RPDM may be of little use as it cannot provide more guidance on fine tuning systems to optimise outcomes.

8.3.1.3 Decision Ladder

The Decision Ladder was developed from Rasmussen's (1983) model of cognitive control which categorises behaviour into three levels of control with a top-level goal-directing behaviour. Skill-based behaviour is automatic and directly interacting with the environment. Rule-based behaviour involves the stored rules and intentions that are activated by cues within the environment. Knowledge-based behaviour involves mental models and analytical problem solving using available information. Decision Ladders present the information, activities and decisions that are involved in making a decision across these three levels of behaviour.

In a Decision Ladder representation of the engine failure on take-off scenario, the primary goal of the scenario was identified: 'to land the aircraft safely as soon as possible with two damaged engines' (see Figure 8.5). Physical environmental cues are the alert to the event. The ECAM/EICAS are also listed as possible alerts, as they have an alerting system when engine parameters are abnormal. The pilot's response to an alert is a skill-based behaviour (Rasmussen, 1983), meaning that it is immediate and automatic. In other words, the pilots know they must respond. The observation of this alert is then noted in the Decision Ladder as a data processing activity. As Decision Ladders aim to capture the functioning of the system, and not just the individual, this can include the pilot's processing of the alerting information but also the flight-deck technologies that are processing the information from the engine sensors to measure the impact of the bird-strike, e.g. engine temperature and vibration. This then provides a 'resultant state of knowledge' within the Decision Ladder. The processing of the information and diagnosis is considered rule-based behaviour, with pilots using their stored rules to respond to alerts in the cockpit and to try to identify what the situation is. The information presented is largely on the flight deck itself and the engine parameters that the pilots must question to assist their diagnosis. The

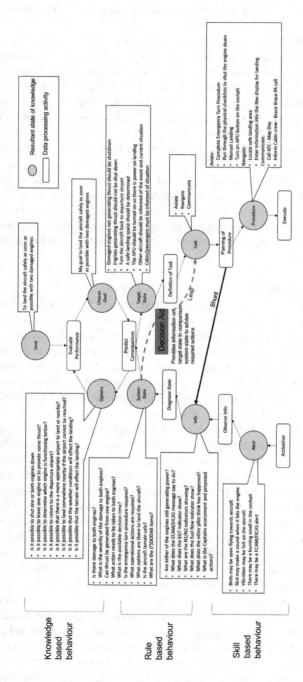

FIGURE 8.5 Decision Ladder showing where the decision aid could provide assistance to the process of managing a dual-engine failure on take-off.

information shared between the two pilots is also used to inform the diagnosis, which is the next data processing activity. Through this activity, the system must determine the damage to both of the engines in order to select the best possible options. This requires interaction between the engine sensors, the engine parameter displays and the pilots' expectations and knowledge of the situation.

At this stage, the environmental conditions are also considered to determine possible landing locations. This information is used to predict the consequences of possible scenarios. Knowledge-based behaviour applies constraints and considers alternatives. As in the RPDM, the expectations of possible outcomes are simulated, except in the Decision Ladder they are not run in serial but concurrently. The different options are shown in Figure 8.5. The high-level system goal guides the option selection.

Shunt shortcuts in Figure 8.5 capture how pilots stay up to date with the constantly updating situation by referring to information on the left-hand side of the ladder to inform their intended tasks and actions on the right side of the ladder. These tasks cover the 'Aviate, Navigate, Communicate' mantra that pilots are trained in. Figure 8.5 also shows how a decision aid could be included within the Decision Ladder to provide a shortcut, from knowledge of the system state to the tasks required to manage the issue. It does so by providing information on the target state. This leap across the system represents the replacement of knowledge-based behaviour with an automated system, as the decision aid can understand what the options are by assessing the state of the system (e.g., both engines are severely damaged and the aircraft needs to be landed as soon as possible).

Not only is the Decision Ladder able to provide insight into autonomous agents within the decision process, it can also suggest how these autonomous agents should function. The questions and responses detailed in the knowledge-based behaviour category on the ladder act as requirements for the information that the decision aid should provide. The pilot is removed from the process of generating different options, predicting consequences, and determining the target state. Furthermore, it could override the need to reference a traditional QRH. The current QRH and checklists were found to be inappropriate for events such as a dual-engine failure by the NTSB (2010) report of the Hudson River incident. Instead, a system that could monitor the continual state of the engine, would be able to inform the pilot of the state of damage to both engines. The decision aid could then inform pilots of the tasks required and time frames for them to be completed in, e.g. to relight the engines or shut them down, and maintain the safety of the flight, action emergency turn procedure, etc. The Decision Ladder is useful here in identifying the role that a decision aid could have within the wider system, as the basis of Decision Ladders is to identify the work required by all actors in a system (Jenkins et al, 2010).

8.3.1.4 Perceptual Cycle Model

The Perceptual Cycle Model (PCM; Neisser, 1976) draws on Schema Theory (Bartlett, 1932) to demonstrate how the environment and context surrounding the decision interact with the cognitive structures and actions of the decision-maker. More information on the PCM approach can be found in Chapter 6 (Section 6.2.1.2).

Following the approach used by Plant and Stanton (2016), the transcripts were coded to the Schema Action World (SAW) taxonomy (Plant & Stanton, 2016) to generate an aggregated PCM. The taxonomy includes 11 World classifications, six Schema classifications and 11 Action classifications, deemed to be fully encompassing of all areas of the PCM (Plant & Stanton 2016). Each of the processes in the PCM was then coded to the different phases of the incident; Pre-incident, Onset of the problem, Immediate actions, Decision-making, Subsequent actions, Incident containment (Plant & Stanton, 2012, 2014).

Figure 8.6 shows the onset of the incident in the 'world', from the physical cues of the bird-strike itself (step 1 in Figure 8.6). This would be subject to a startle and/or surprise effect, reflecting an insufficient schema (2) as pilots do not initially know what is happening due to the unanticipated and unusual nature of the event. The pilot's initial priority to fly the aircraft (3) is captured as the first 'action' in the PCM and it is still considered to be within the initial onset of the incident as the pilot is not initially making any decisions. The remainder of the events follow the Schema Action World themes in numerical order.

Like the Decision Ladder, the PCM also considers the broader system in which decisions occur, but it considers this in relation to the schemata of the individual and how the world, and actions within it, influence the cognitive processing of the individual. Schemata are anticipatory and foresee information in the environment to direct appropriate action. Knowledge structures comprising the schemata can also be updated to guide the exploration and interpretation of future information. The RPDM focuses on the individual's perspective, but it does not account for the interactional nature of the environment and the impact it has on shaping behaviour (Lipshitz & Shail, 1997; Plant & Stanton, 2014). There are no feedback or feed-forward loops to capture how an individual communicates with their environment (Plant & Stanton, 2014). This is also evident in the starting point of decision-making. Both the PCM and Decision Ladder start with the initial event in the external environment and how it first presents itself to the decision-maker, e.g. a physical cue (bang/vibration/smell) and/or the engine instruments on the flight deck. Conversely, the RPDM starts from the point at which the event is experienced by the decision-maker and proceeds to assess their cognitive processing of this experience.

The interactional nature of the individual with the environment within the PCM means that it requires information to be accurate and up to date. Real-time information on the state of the engine would assist in reducing the ambiguity in the decision-making process. The introduction of a decision assistant that could provide such information is shown in Figure 8.6 as assisting at both the immediate response stage and the decision-making stage. An aid that could provide real-time information on the state of damage to the engine itself would inform the pilot's initial expectations of the event (steps 5 to 6 on Figure 8.6) and guide the pilot's response.

As option generation occurs in parallel, referring back to the information on the engine state throughout the decision-making process between steps 11 and 12 would provide a more accurate representation of the event in the pilot's schema and therefore an improved chance of an accurate diagnosis in the following stages of decision-making. This varies from the RPDM which suggests the importance of an aid to help

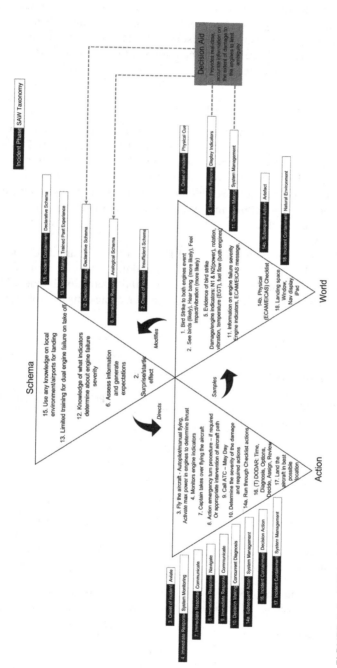

FIGURE 8.6 PCM of the dual-engine failure scenario showing where the decision aid could provide assistance to the pilot on the flight deck.

recognise the situation at the very beginning of the process due to its serial decision-making process. Furthermore, an awareness of the changing situation with updated information would also allow for more optimal decisions, rather than satisfactory ones. Applying the PCM can therefore assist in understanding how a decision aid will interact with the cognitive processing of the pilot and the wider environment, as this changes across the decision event.

8.3.2 COMPARING DECISION MODELS

The purpose of modelling decision-making is to understand where it can be improved or guided using decision aids (Banks et al, 2020a; Dorneich et al, 2017; Mosier & Manzey, 2019; Mosier & Skitka, 1996; Parnell et al, 2021a, 2021b; Simpson, 2001). Three different decision models have been applied to pilots' responses to a dual-engine failure on take-off event. The different decision models provide some complementary and some varying insights into how pilot decision-making can be supported in the event of a dual-engine failure on take-off. A summary of the differences is shown in Table 8.6.

As depicted in Table 8.6, the RPDM and Decision Ladders are divergent in their approaches, yet the PCM falls somewhere in the middle of the two, as it can account for the cognitive processing of the individual and the interaction of the wider systemic elements. The ability to account for the wider context has enabled the PCM

TABLE 8.6
Summary of Key Differences Across the Decision Models

Aspect of decision-making	RPDM	PCM	Decision Ladder
Decision must be justifiable	No *Non-optimum decision-making*	Yes	Yes
Options are generated in parallel and compared	No *Options considered in serial*	Yes	Yes
Focuses on the cognitive processing of an individual decision-maker	Yes	Yes	No *Focuses on tasks in the system*
Accounts for previous experience of the decision-maker	Yes	Yes	No *Focuses on tasks in the system*
Accounts for the interaction of other actors in the decision-making process	No *Focuses on the individual*	Yes	Yes
Starts with initial event in the environment	No *Starts with the experience of the event by the individual*	Yes	Yes

and Decision Ladders to be utilised for exploring decision-making with autonomous agents (Banks et al, 2018a; Revell et al, 2021; Roth et al, 2019). To enable successful human–autonomous interactions, models that incorporate autonomous agents and their interactions with the environment will be necessary. The RPDM may be somewhat outdated in this respect, it relies on the recognition for events based on past experiences, yet many of the experiences that autonomous agents will bring will be novel and there will be a steep learning curve to understand how their decision-making processes relate to our own, if they do at all.

The findings from this research suggest that models that can account for the role of the environment will be able to provide in providing more detailed information on the systems dynamic requirements. They can help to provide updates and real-time feedback. The Decision Ladder approach is particularly useful in scoping out the information that a decision assistant needs to provide in order to replace human knowledge. Therefore, this method is advocated for those considering the design of an automated decision aid. The PCM, however, is particularly useful at presenting how a decision aid will be integrated within a certain scenario and how it will shape the experience of the individual, as well as their interactions with the environment. Therefore, this method is advocated in the review of a decision aid after its initial design to understand how it may interact within a system. This could allow any unforeseen and unintentional interactions to be reviewed early on in the design process.

Conversely, the RPDM shows value in the initial alerting of the need for the decision and the interpretation of the situation but not how this can relate back to the environment. Its use may therefore be more limited to the design of alerts once systems have failed, in order to inform individuals of the situation and guide their response. Table 8.7 summarises the different insights into future measures from the different models in relation to the dual-engine failure scenario.

TABLE 8.7
Summary of the Decision Model Recommendations

Decision Model	Recommendations
RPDM	• Allows early recognition of the full severity of the event, i.e. cues to the damage to both engines at the same time to give accurate expectations of future actions • Presents information in a familiar format so that they can recognise it as a dual-engine engine failure
PCM	• Reduces ambiguity in the state of damage to the engine to update the pilot's schema • Provides real-time, accurate data that can be referred to throughout the decision-making process
Decision Ladder	• Provides information in current system state and target system state by processing the options available and presenting the best course of action for pilots to action • Short-cuts the decision-making and limits the need for knowledge-based behaviour from the pilot

8.4   CONCLUSIONS

This chapter has illustrated how we can model both the system as a whole and the interactions between different actors with OESDs. It has also presented different tools for modelling specific user behaviours in relation to decision-making. Both of these approaches can show the implications of introducing new technologies to current work systems. OESDs show how new technologies may change and update the interactions and tasks conducted by different actors. This is useful to model, in order to review how a new technology may change work practices and highlight any previously unseen consequences or adverse events. OESDs can also allow the operational loadings that tasks require of actors in the diagram to be calculated. This is particularly useful when looking to implement automation within current systems, to review how the task loadings differ with increased levels of automation.

The user modelling approaches that were presented focus on human decision-making. The different approaches highlight how different ways of modelling can lead to different outputs and also different design recommendations. Each of the models presented could capture the impact of a decision aid to the decision process. It is important to understand the differences in the approaches as there are numerous decision models available in the literature. Within our work we emphasise the importance of the environment and context on decision-making. Comparisons between the decision models (RPDM, Decision Ladders and PCM) found that the PCM was best able to account for the role of environment as well as individual in relation to the decisions made.

These modelling techniques are useful in anticipating how new technologies will integrate within current systems. They can help identify the need for new procedures as well as the potential for possible implications and adverse outcomes. The modelling stage is low in cost but can add much value, allowing adjustments to be made before the full systems and technologies are created. Once modelled, the next stage is to evaluate and assess the performance of the system. The next chapter covers the evaluation stages.

9 Design Evaluation

9.1 INTRODUCTION

Design evaluation is considered to be the final stage of the design lifecycle and it is critical to determining if the design can, and should, be integrated into the future system. Design evaluation should be conducted throughout system development, and the form this evaluation takes will be dependent upon the degree to which the design is developed. This aligns with the reiterative nature of the design lifecycle, whereby improvements can be made throughout development and by engaging end-users at critical evaluation points. This chapter covers three evaluation scenarios of potential future flight system technologies that can occur at various stages of the design development. The benefits of evaluation in system design are highlighted, both in terms of efficiency and in terms of the role it plays in fostering innovation in design. Examples from these different real-world design developments are discussed from pre-development design through means of wire-frame evaluations (Section 2), functional simulation evaluation (Section 3), and finally usability simulation evaluations (Section 4).

9.2 WIRE-FRAME EVALUATIONS

Wire-frame evaluations are rudimentary mock-ups of the future system that may be used in various capacities in all stages of the design life cycle. For example, in 'Stage 1: Design' of the design process in Chapter 5, mock-ups may be generated by participants as part of the Design with Intent (DwI) method (Lockton et al, 2010). These would likely be annotated drawings of how a new technology would function. Modelling within Stage 2 would result in more sophisticated versions of these designs, however, they would still be annotated images or videos demonstrating the simplistic functions of the design. This would allow for evaluation of the potential technology without the exorbitant financial and time investments associated with building a prototype.

The application of heuristic evaluation methods enables practical consideration of different design options without being resource intensive and requiring functional end-product systems.

DOI: 10.1201/9781003384465-9

TABLE 9.1
Nielsen's (1994) Usability Heuristics

Heuristic	Description
Visibility of system status	Always keep users informed about what is going on, through appropriate feedback within reasonable time. For example, if a system operation will take some time, give an indication of how long and how much is complete
Match between system and the real world	The system should speak the user's language, with words, phrases and concepts familiar to the user, rather than system-oriented terms. Follow real-world conventions, making information appear in natural and logical order
User control and freedom	Users often choose system functions by mistake and need a clearly marked 'emergency exit' to leave the unwanted state without having to go through an extended dialog. Support undo and redo
Consistency and standards	Users should not have to wonder whether words, situations or actions mean the same thing in different contexts. Follow platform conventions and accepted standards
Error prevention	Make it difficult to make errors. Even better than good error messages is a careful design that prevents a problem from occurring in the first place
Recognition rather than recall	Make objects, actions and options visible. The user should not have to remember information from one part of the dialog to another. Instructions for use of the system should be visible or easily retrievable whenever appropriate
Flexibility and efficiency of use	Allow users to tailor frequent actions. Accelerators – unseen by the novice user – may often speed up the interaction for the expert user to such an extent that the system can cater to both inexperienced and experienced users
Aesthetic and minimalist design	Dialogs should not contain information that is irrelevant or rarely needed. Every extra unit of information in a dialog competes with the relevant units of information and diminishes their relative visibility
Help users recognise, diagnose and recover from errors	Error messages should be expressed in plain language (no codes), precisely indicate the problem and constructively suggest a solution
Help and documentation	Few systems can be used with no instructions so it may be necessary to provide help and documentation. Any such information should be easy to search, focused on the user's task, list concrete steps to be carried out and not be too large

Heuristics provide generic usability criteria to assess the interfaces against. Nielsen's 10 Usability Heuristics (Nielsen, 1994) are the most widely used and accepted heuristics for interface evaluation (see Table 9.1) and were therefore applied to our work. Conducting a heuristic evaluation is a relatively quick, inexpensive and easy method for evaluating multiple designs before they go into production. The evaluations can be done on rudimentary wire-frame mock-ups of the displays (e.g., via PowerPoint presentations or simple drawings). Nielsen (1994) argues that the ease of use of any system or design must be paramount as difficult technology defeats the purpose or goal of improving the system. Therefore, the evaluator rates each heuristic in terms of the severity of any usability problems.

Nielsen recommends around five respondents, as a saturation point is achieved and minimal gains in information are found in larger group sizes. It is acknowledged that there will always be usability errors that are easily identified, and therefore will be identified by nearly all evaluators; however, some less obvious usability issues may only be identified by a few respondents.

Heuristic evaluations can be conducted with Human Factors professionals, who have an understanding of interface usability, as well as the intended user group who have an understanding of the context in which the interface is to be used. Evaluations from both groups can be complimentary and enhance the validity of the findings. Therefore, this was targeted in this work. Presented here is an overview of a heuristic evaluation that was conducted for three wire-frame designs of a proposed digital indicator of engine status for warm-up and cool-down.

9.2.1 Case Study: Warm-up/Cool-down Indicator

The purpose of the warm-up/cool-down indicator (WUCI) is to provide pilots with real-time information regarding the recommended warm-up time of the engine pre-flight, as well as encouraging appropriate cool-down times post flight. The warm-up indicator would allow for optimisation of engine start up time, permitting single-engine taxi where possible and indicating when to turn on the second engine, in accordance with engine state and environmental conditions. Similarly, the cool-down indicator aims to show the recommended and minimum engine cool-down time in order for it reach a required thermal state. These types of indications are novel, and while not safety critical, can have a significant impact on the long-term preservation of the engines.

Prior to the heuristic evaluation, initial designs were developed through the use of the DwI workshops with pilots (Lockton et al, 2010). This led to the development of four different indictor designs, which were then subjected to heuristic evaluation. Evaluation of the designs was done by Human Factors experts to assess the usability of the interactions with the system, and with commercial airline pilots to assess the integration and functionality of the system by the intended end-user group. Both groups used Nielsen's 10 Usability Heuristics (Nielsen, 1994; see Table 9.1) to evaluate the designs.

9.2.1.1 Participants

Three males and three females responded to the survey, with an average age of 28.2 years ($SD = 4.1$) and 6.25 years' experience (on average, $SD = 3.47$) within the domain of Human Factors. While two had previous experience in research within the field of aviation, no Human Factors experts were qualified pilots.

Six males responded to the pilot online survey, with an average age of 44.00 years ($SD = 13.27$). Pilots all held their APTL, having held their licences for 16.83 years ($SD = 16.06$, *range:* 5–44 years). Within their current roles, two participants held the position of Captain (one being "Line Training Captain"), two were Senior First Officers, and the remaining two were First Officers. Pilots either currently, or had previously held, positions with airlines such as Air UK, British Airways, Eastern Airways, Germanwings, Lufthansa, Ryanair, Tui and Virgin Atlantic.

9.2.1.2 Designs

Following the initial design workshops, a researcher in the team developed wire-frames of the designs (see Figure 9.1).

Design 1 (Figure 9.1a) included different designs for the warm-up and cool-down phases. Warm-up was indicated by an increasing colour band around the exterior of the N1 engine gauge; the colour of the band changed from amber to green, as the warm-up increased with green indicating the engine was warmed and ready for use. The cool-down indicator appeared initially as a timer, counting down the estimated time until the engine would be sufficiently cooled. An amber arrow would appear indicating which engine the countdown was referring to, with the arrow changing to green when the engine was ready for shutdown.

Design 2 (Figure 9.1b) included the same integration for both warm-up and cool-down. Appearing on the Systems Display screen, at the relevant stage of the flight, a timer appears adjacent to the fuel usage indicating the countdown for the engine. The counter appears in white with a pulsating outer box while counting down, which changes to amber in the final 30 seconds to countdown.

Design 3 (Figure 9.1c) used essentially the same design for warm-up and cool-down. This indicator appears on the standard engine display screen in the accompanying message panels. A countdown timer is integrated into the right side, with red text counting down the time until an engine is warmed or cooled and changing to green when ready. The point of difference between the warm-up and cool-down phases is the integration of the engine status into the pre-flight configuration check; with a blue message (consistent with the current display) appearing to indicate if an engine has not reached the warm-up thresholds.

Design 4 (Figure 9.1d) included different designs for the warm-up and cool-down phases. The warm-up indication was similar to the display for Design 3, with a countdown timer appearing in the text panel of the standard engine display. Colour is integrated into the message to indicate when the engine is warmed sufficiently. For cool-down, a green light would appear in the centre engine console, below the glare-shield display to indicate when the engine is ready for shutdown. There is no countdown timer, only the light to indicate when ready.

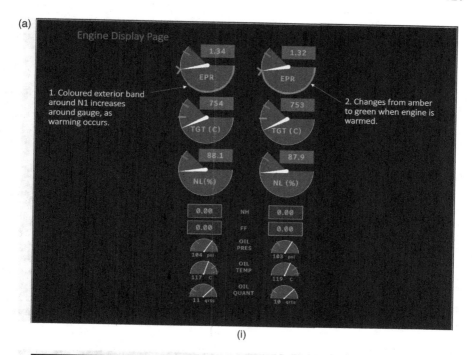

(i)

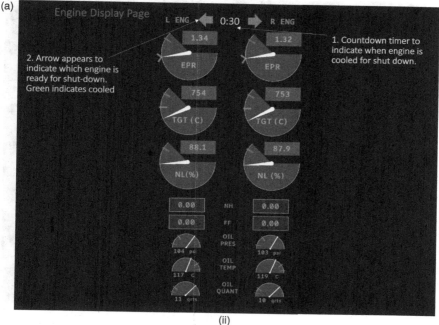

(ii)

FIGURE 9.1 Wire-frame designs for warm-up/cool-down indicator generated from user-led design workshops.

(b)

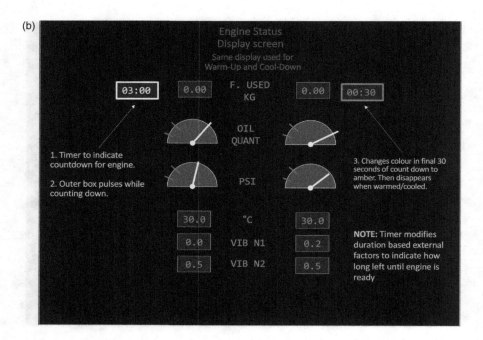

1. Timer to indicate countdown for engine.

2. Outer box pulses while counting down.

3. Changes colour in final 30 seconds of count down to amber. Then disappears when warmed/cooled.

NOTE: Timer modifies duration based external factors to indicate how long left until engine is ready

(c)

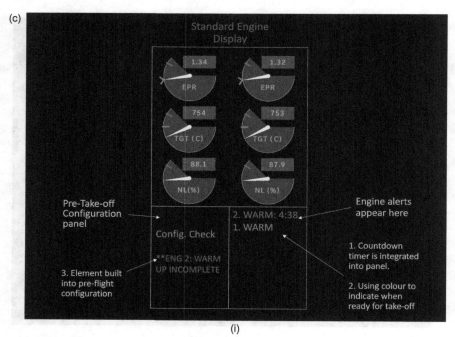

Pre-Take-off Configuration panel

3. Element built into pre-flight configuration

Engine alerts appear here

1. Countdown timer is integrated into panel.

2. Using colour to indicate when ready for take-off

(i)

FIGURE 9.1   Continued

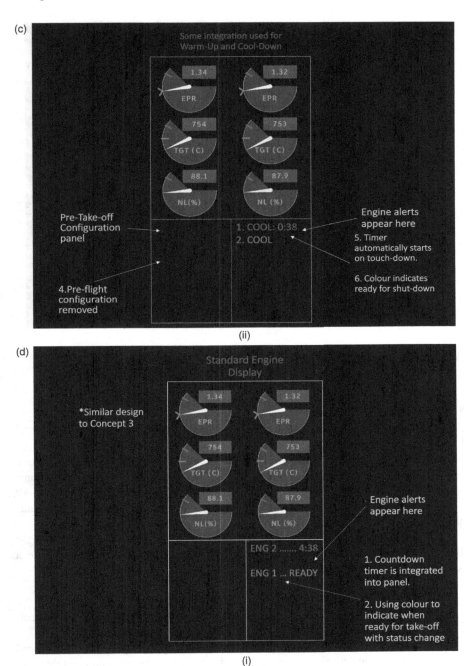

FIGURE 9.1 Continued

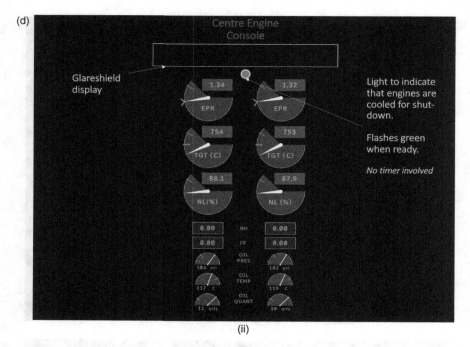

FIGURE 9.1   Continued

9.2.1.3   Heuristic Evaluation

The evaluation was presented to participants in an online survey. Nielsen (1992, 1994) has highlighted the benefits of individual assessments for heuristic evaluation, as it allows users to consider their own experience and not be influenced by others. Group discussion after the evaluation can be beneficial to understand the evaluations, however in this case it was not possible due to restrictions imposed by the COVID-19 pandemic. Each design was presented to participants in an instructional video outlining the wire-frame designs.

For each design, participants are asked to apply a rating indicating if they thought there was a 'usability problem' associated with each heuristic. Ratings were on a scale of 0–4 as outlined here:

- 0 None – I don't agree that this is a usability problem at all;
- 1 Cosmetic problem – need not be fixed unless extra time is available on project;
- 2 Minor usability problem – fixing this should be given low priority;
- 3 Major usability problem – important to fix, so should be given high priority; and
- 4 Usability catastrophe – imperative to fix this before product can be released.

Additional comments and feedback as it pertained to each heuristic (in lieu of the wider group discussion) were captured in open-field text responses alongside their rating. Finally, respondents indicated their preferred design.

9.2.1.4   Results

Ratings from the two participant groups (Human Factors experts and pilots) were compared and there were no overwhelming differences between the cohorts' data. Therefore the data are considered in combination. That said, one point of difference was that when a rating of 4 (indicating a 'Usability catastrophe – imperative to fix this before product can be released'), these response ratings only came from pilots. This was considered to be based on their knowledge and insight into how the indicator would integrate into the specific flight stages and the likelihood of severe impacts.

Table 9.2 displays the most common rating for each heuristic across each design, with a summed total heuristic score. As indicated, Design 1 had the least usability issues, with the most common response for each heuristic always 0. Design 4 was evaluated as having the most issues with usability, including two instances of major usability concerns.

The final question related to the preferred design for the indictor. Participants were asked to rank each design from most to least preferred; with outcomes depicted in

TABLE 9.2
Most Popular Usability Ranking by Design

Heuristic	Design 1	Design 2	Design 3	Design 4
1. Visibility	0	0	2	3*
2. Real-world match	0	0	0	0
3. User control	0	1*	0	1*
4. Consistency and standards	0	0	0	0
5. Error prevention	0	0	0	3*
6. Recognition	0	0	0	1*
7. Flexibility	0	0	0	0
8. Aesthetic	0	0	2*	1*
9. Error recovery	0	0	0	0
10. Help	0	1*	0	0
Total	**0**	**1**	**4**	**9**

Note. Values indicate usability ratings: 0-None, 1-Cosmetic, 2-Minor, and 3-Major; *indicates this value was tied with '0' as the most popular response.

TABLE 9.3
Participant Ranking of Designs from Most to Least Preferred

Participants	Participants Ranking of Designs from Most to Least Preferred			
	Most Preferred	Second Most	Third Most	Least Preferred
Human Factors experts	1, 3	–	2	4
Pilots	1	3, 4	–	2
All Combined	1	3	2	4

Note. When two designs were ranked equally, such as when for HF experts ranked first, both appear.

Table 9.3. When considered separate from one another, the pilots overly preferred the first design, and the Human Factors experts were equal in their preference for Designs 1 and 3. Overall, the preferred design was Design 1.

9.2.1.4.1   Usability and Acceptance of Preferred Design

While the results of the heuristic evaluation did indicate a preferred design, it was necessary to conduct subsequent evaluations of the design for usability and suitability. A follow-up pilot study was completed for usability and acceptance of the preferred design (Design 1, Figure 9.1a). This was presented as an online survey and shared to pilots who had previously provided their details for research participation.

Twelve male pilots completed an online survey, with an average age of 47.50 years (SD = 12.44). Ten pilots held their APTL, with the remaining two pilots holding a CPL. Participants held their licence for an average of 20.76 years (SD = 13.32, *range*: 4–40 years), with an average of 10 291.03 accrued flying hours (SD = 8 499.93).

The survey consisted of two widely used subjective questionnaires. First, the System Usability Scale (SUS; Brooke, 1996) is comprised of 10 questions, and is used to calculate the overall usability of an interface's design. Second, the Acceptance Scale (Van Der Laan et al, 1997) is comprised of nine questions, used to assess acceptance of a system on two dimensions: usefulness and satisfising. Scores are rated on a scale from –2 to +2, with scores over 0 considered to be highly accepted by the user.

Given the small sample size, only general comments can be made on the evaluation of the designs. The design was scored as 'above average' on usability, with a SUS (Brooke, 1996) score of 75.21. The participants rated the design highly on both usefulness and satisfising scales of Van Der Laan et al.'s (1997) Acceptance Scale (M = 1.49 and M = 1.27, respectively).

9.2.1.4.2   Comments and Recommendations

Throughout the heuristic evaluation and usability and acceptance evaluations, participants were also asked to provide any further comments on the design, with most comments positively rated. The consensus was that the design was simple and that it met the needs of the pilots, as they were not concerned with additional engine information, only that the engine was ready for use. Some comments were made indicating that the warm-up design could be confusing in a B737 display, therefore further consultation with design teams would be required.

The initial findings indicate that the chosen design would be suitable, however there needs to be further testing and wider participant sampling before further recommendations can be made. Especially given the small sample size for the Acceptance and Usability Scales, this would need to be replicated with more participants and to have the comparison with the current interface, with the addition of workload parameters to determine if the instrumentation is more effective than current procedures. The current evaluations demonstrate the suitability of the online format for these evaluations, and so this platform can be used again in the future.

One limitation of this evaluation was that it did not differentiate between the two functions (warm-up or cool-down) when evaluating each design. Therefore, it would be helpful to understand if the end-user would be interested in using the same design for the warm-up and cool-down. Specifically, there does not seem to be a preference for having a separate or combined indicator; while Design 1 was the preferred (separate designs), Design 3 was also rated highly and Design 2 was close in terms of usability (both of which use a combined design). Concerns were raised over the overload that can occur at taxi, the 'simple' timer for cool-down would be preferred, rather than adopting the same arc-design as used in warm-up.

In conclusion, the results of the wire-frame evaluation suggest that at this stage, Design 1 shows promise for integration into the flight in terms of its usability and acceptance levels. However, further validation of the design would be required and this would be best achieved in a flight simulator configuration whereby the design can be integrated into the flight deck environment.

9.3 SIMULATION EVALUATIONS

This section presents the second part to the evaluation of new design concepts. Once a concept has been proposed and selected from the wire-frames stage, the next step is to create the functional design for evaluation in a simulated environment. With the development of advanced flight simulators, examples of new technologies can be evaluated through user testing with the intended end-user sample. This section highlights the benefits of simulator studies to evaluate the usability of the design within the context of its intended use.

As with wire-frame evaluations, simulator evaluations allow for an assessment of the proposed future system, without the associated expenses of full prototype development. While the designs used in simulation are more sophisticated than wire-frames, they are still models of the intended future system and may require an experimenter to control certain functions that are perceived to the participant as being automatic. Simulator evaluations allow for more objective evaluations, as one can investigate how the end-user would interact with the future system in a model of its intended use. It can also allow for the testing of multiple proposed designs alongside existing configurations, to determine its suitability before integration. A major concern of all simulated environments, however, is the issue of ecological validity, or the extent to which research outcomes can be generalised to the real world. Within the context of aviation, validity of flight simulation is already well established, and simulation forms an essential component of pilot training programmes and research and development (Rolfe & Staples, 1988).

Whilst in industrial research and development, test pilots are typically used as participants, they do not necessarily represent the end-user population (i.e., typical line pilots). Test pilots are a highly specialised cohort of individuals who complete flight test techniques (e.g., focusing on specific manoeuvres) to enable the design of a system to be evaluated. While critical in design, development and some evaluations, they may not engage with the system with the same intentions as an average line pilot during typical day-to-day operations. To fully understand the interaction between the pilot and any new technology on the flight deck, it is therefore important that research be conducted with a representative pilot cohort, as is the case in the following evaluation case study.

9.3.1 CASE STUDY: OIL STARVATION AVOIDANCE (OSA)

An oil starvation avoidance feature is proposed as a tool to provide the pilot with advanced warning of a possible oil starvation event resulting from an oil leak. The feature aims to reduce disruption to a flight by providing the flight crew with more time to make informed decisions about appropriate courses of action. Further information relating to this case study can be found in Chapter 6 (Section 2.2).

The presentation of engine system parameters has remained largely unchanged for a number of decades, with analogue displays simply being replaced by digital equivalents (Harris, 2011). This means that in current practice, pilots only become aware of abnormal system parameters when threshold limits have been met via their on-board alerting systems. At the point of notification, pilots have few options available to them and they must either throttle back or shut down the engine completely, to prevent the engines from reaching oil starvation (Australian Transport Safety Board [ATSB], 2012, 2017). However, the proposed OSA feature shall:

1. Improve pilot performance in dealing with abnormal engine system parameters;
2. Improve engine utilisation; and
3. Reduce the number of in-flight shutdowns, air turn backs and diversions.

Previous work has used end-user involvement to provide insight into the design requirements surrounding a novel OSA feature (see Chapter 7, Section 2.2). Two stages of evaluation were conducted with the OSA feature once initial designs were generated. This involved down-selection using heuristic evaluation and feedback, plus testing of the final proposed design in the flight simulator. Of interest here is the testing of the design within a flight simulator.

9.3.1.1 Experimental Scenario

Following DwI workshops and heuristic evaluations, a single design for an OSA alert was developed. Flight simulator testing was conducted on the Future Systems Simulator (FSS) at Cranfield University.

An experimental design scenario was designed within the simulator platform that allowed for the design of interest to be tested. This scenario required pilots to take off from an airport (with no maintenance option) and then fly to a destination airport (where there was maintenance). After reaching a predetermined point in the flight, the oil indicators started to show decreasing oil levels. In one condition (i.e., the baseline or control condition) the pilots would not receive any indication of the oil level until in the cruise phase of the flight, which is consistent with current flight deck technology (OSA 1). In the second (experimental) condition the oil alert (OSA 2) would be implemented and it would provide an advanced warning. All participants experienced both conditions, in a random order. The impact of the newly designed oil alert feature on pilot decision-making and response was the focus of the simulator study.

9.3.1.2 Participants

Airline pilots with experience of twin-engine aircrafts were recruited to participate. Using professional networking sites and databases generated from previous research trials, 81 pilots were initially recruited. From the initial 81, 12 pilots (males $n = 10$) were selected to participate in the flight trials, where they were then arranged by the research team into pairs based on their availability and experience. Due to subtle differences based on training within airlines and roles, it was important that the pilots were paired together by a researcher with knowledge of their experience. It was important that the pairs were not disproportionally different from the others in terms of experience, and that the pilots within the pairs had similar aircraft type and similar employer experience. Participants were categorised across four age categories: 31–40 years ($n = 7$); 41–50 years ($n = 2$); 51–60 years ($n = 1$); and 60 years+ ($n = 2$). Participants had an average of 9392 flight hours logged ($SD = 7809$) and 14.22 years experience in commercial flight ($SD = 9.92$).

9.3.1.3 Results

The decisions made within each of the six trials are presented in Table 9.4. There was a technical malfunction within Trial 1 and therefore results from this trial for the OSA 1 condition could not be included. All trials in the OSA 1 condition resulted in an in-flight engine shutdown, apart from one trial (Trial 6) where the pilots chose to divert from the planned destination. Trial 6 continued on to their destination.

In the OSA 2 condition, where the OSA feature was active, all bar one trial (Trial 6) continued flying with one engine. Trial 6 chose to shut down the left engine but then elected to perform an Air Turn Back (ATB). All trials ultimately opted to perform an ATB, apart from Trial 3, who choose to divert, as indicated in Table 9.4. As nearly all trials opted to perform the ATB, it suggests that the OSA feature does lead to improved decision-making.

TABLE 9.4
OSA Scenario Outcomes

| TRIAL | OSA 1 – Without OSA | | OSA 2 – With OSA | |
	Engine Mgt.	Flight Mgt.	Engine Mgt.	Flight Mgt.
TRIAL 1	N/A	N/A	Continue with 2 engines	ATB
TRIAL 2	RH ENG shutdown	Divert – RKPK	Continue with 2 engines	ATB
TRIAL 3	RH ENG shutdown	Divert – RJOA	Continue with 2 engines	Divert – RKTN
TRIAL 4	RH ENG shutdown	Divert – RKPK	Continue with 2 engines	ATB
TRIAL 5	RH ENG shutdown	Divert – RKPK	Continue with 2 engines	ATB
TRIAL 6	RH ENG shutdown	Continue – RKSI	LH ENG shutdown	ATB

TABLE 9.5

Breakdown of the Trials Across the Different Stages of the OSA 1 Flight Scenario

Incident Phase	TRIAL 2	TRIAL 3	TRIAL 4	TRIAL 5	TRIAL 6
Pre-incident	Climb phase	Cruise phase		Climb phase	Cruise phase
Onset of problem	Secondary engine indicator	Master Caution		Secondary engine indicator	Master Caution
Immediate Actions	Run checklist & call ATC – Pan/Mayday THEN Descend to lower altitude			Descend to lower altitude THEN Run checklist & call ATC – Pan/Mayday	Run checklist & call ATC – Pan/Mayday THEN Descend to lower altitude
Decision-Making	Engine shutdown required				
Subsequent Actions	Identify suitable place to land. Decision to divert				Continue to destination
Incident containment	Brief for landing				Brief for landing

TABLE 9.6

Breakdown of the Trials Across the Different Stages of the OSA 2 Flight Scenario

Incident Phase	TRIAL 2	TRIAL 3	TRIAL 4	TRIAL 5	TRIAL 6
Pre-incident	Approaching top of climb				
Onset of problem	ECAM message				
Immediate Actions	Run through checklist & OSA application				
Decision Making	Decision to return to departure airport				Decision to divert
Subsequent Actions	Monitor engine instruments				Execute engine shutdown
Incident containment	Brief for landing				Brief for landing

Table 9.5 shows that although the decision to shut down the engine was unanimous across all trials in OSA 1, the steps preceding this decision varied, as did the subsequent actions and incident containment.

In comparison, there was considerably less diversity in the responses to the oil leak scenario when the OSA feature was present. Table 9.6 shows the processes for each trial across the different stages of the OSA 2 scenario. This shows the only deviation was Trial 6 and their decision to divert and execute an engine shutdown. However, the process of detecting the problem and immediate actions taken were the same across all trials.

The pilots' responses were further reviewed to study how the pilots responded to the OSA feature. All pilots detected the oil leak via the OSA feature on the ECAM/EICAS system. This allowed all trials to identify and diagnose the issue within the climb phase. The OSA system successfully triggered their prior knowledge of the event and the processes that they have been trained in. All pilots could then successfully respond using the checklist and ECAM/EICAS messages.

Trial 6 followed a different decision-making process. They identified that an engine failure on a go around or approach is more difficult to deal with, which would make the option of performing an Air Turn Back (as all other trials performed) more difficult. Therefore, they felt continuing to a diversion airport was a better option.

9.3.1.3.1 *Comments and Recommendations*

Reviewing the decision-making of the pilots who experienced the baseline trial without the OSA and feature (OSA 1) and the scenario with the OSA feature (OSA 2) suggests that the feature allows for a quicker response by the pilots, which gives them more options for controlling the flight and preserving the engine. It also created less diversity in pilot decision-making, with the majority choosing not to shut the engine down.

Involving airline pilots within the evaluation process has been highly useful in understanding the utility and usability of the OSA design concept. This is because they are able to provide more insight into the use of the equipment that non-users or design experts may fail to understand. However, a limitation of this study is the small sample size, which was primarily due to recruitment barriers. Therefore, when possible, and user-specific knowledge is not required, it may be possible to use members of the general population for testing.

9.4   SIMULATION ACROSS CONTEXT OF USE

As identified in Chapter 4, the context of use is intrinsically linked to the determined usability of designs. Within cockpit design, it is vitally important to consider all contexts in which designs must work effectively. With the move towards touchscreens in aircraft cockpits a key contextual factor in usability is turbulent conditions. While the previous section highlighted how simulators can be useful in evaluating functional designs, they can also be useful in simulating different contexts of use. Presented here is a two-stage simulator study that investigates the usability of touchscreen interactions across turbulent conditions.

9.4.1   CASE STUDY: FLIGHT-DECK TOUCHSCREENS IN TURBULENT CONDITIONS

As the plans for an integrated flight deck continue to advance, further research is required to optimise performance and minimise risk and errors. Some important research has addressed the optimum physical aspects of the touchscreen design to reduce these errors, such as button size (Bender, 1999; Dodd et al, 2014; Hoffman & Sheikh, 1994), screen selection (Dodd et al, 2014; Happich, 2015) and auditory (Bender, 1999) or haptic feedback (Graham-Rowe, 2011), with some potential improvements identified. However, most of this research supporting the use of touchscreens in the aviation domain has been conducted in static conditions despite the reported impacts of turbulence on performance and usability (Coutts et al, 2019). Turbulence is characterised by inadvertent movement in all axes of motion and can therefore impact an individual's ability to move freely. Orphanides and Nam (2017) identified that touchscreen technology utilisation is heavily dependent on the environmental conditions in which it is used. As turbulence on flights is predicted to become increasingly prevalent with future predicted climate change (Williams, 2017), it is an important factor to consider in the development of future flight deck technology.

9.4.1.1 Task-specific Testing

Efforts are being made, in commercial aviation, to integrate all flight-deck functions into an open touchscreen-based graphical display. This means that the traditional, mechanical, units of input will need to be integrated into touchscreens. A potential integration of a pan task on the flight deck could be in the format of a graphical interface, like a car's navigational display, where a user may wish to move the map to view surrounding features. Alternatively, like arranging 'apps' on a display, pilots may wish to bring to the forefront a specific menu or 'task' list; doing so by a pan gesture. This represents a potential enhancement of flight-deck tasks through touchscreen integration. Another, so far neglected, area of research is the testing of number entry in a touchscreen format, under control and turbulent conditions, in commercial aviation. Number entry forms the basis of many in-flight activities, such as selecting the heading, altitude or airspeed in the Flight Director, and modifications to these may be required if unexpected or continuous turbulence is encountered. Currently this is achieved through a rotational mechanical dial, however integration into a touchscreen display is proposed. Therefore, the current study will assess common touchscreen gestures (e.g., tap, tap-and-hold, swipe, drag) with the goal of inputting a three-digit number. As this task is currently completed by pilots as part of their regular flight operations; the touchscreen gestures will be compared to the traditional configuration of a mechanical dial, to determine if the touchscreen improves performance.

Successful touchscreen integration into the flight deck should ensure tasks can be completed as quickly and accurately as possible, without increasing workload or impacting comfort. It was hypothesised that an increase in turbulence will result in longer task completion times and additional touchscreen interactions. It was also predicted that turbulence will negatively impact the workload and usability of the tasks.

9.4.1.1.1 Panning Tasks

The goal of the pan task was to move a 2.5 cm diameter solid blue circle into a larger 3.5 cm grey circle ('gate'). The three task variations (shown in Figure 9.2) were: (a) tap directional arrows; (b) tap-and-hold directional arrows; and (c) drag the circle itself into the target gate. The main difference between 'tap' and 'tap-and-hold' was that for tap only, a single press would move the circle 0.9 cm, whereas for tap-and-hold, if the participant were to press and hold the arrow the circle would continue to move at a constant rate (19 cm/second) until the arrow was released. These tasks were selected as they represent different ways a pilot may be required to move icons around a map.

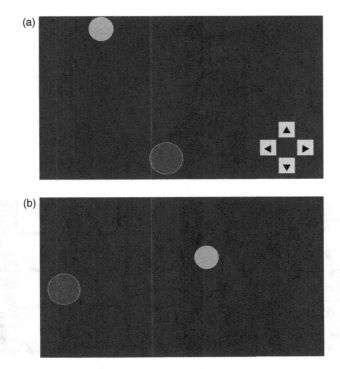

FIGURE 9.2 Sample screenshots of pan tasks.

9.4.1.1.2 Number Entry Tasks

For the number entry tasks, participants were required to match a three-digit target number using different inputs for each of the eight task variations. The tasks (shown in Figure 9.3a–h) were:

a. mechanical dial;
b. number entry keypad;
c. arrows aside, tap three pairs of directional arrows, above and below the numbers to change digits individually;
d. arrows below, tap three pairs of directional arrows, below the numbers to change digits individually;
e. arrows, tap-and-hold single set of directional arrows to scroll through a three-digit number sequentially;
f. vertical rectangular bars, swipe up and down in bars located below the numbers to change digits individually;
g. vertical sliders, drag circles located on each step of a vertical ladder up and down to change digits individually; and
h. clock-face, tap and drag a circle around the clock-face design to change each digit individually.

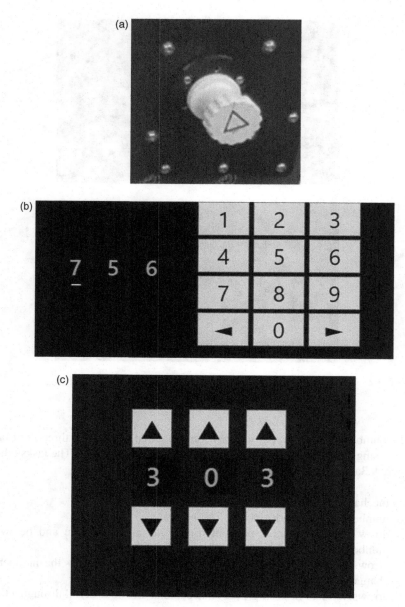

FIGURE 9.3 Visual displays for number entry tasks with required screen interaction skill.

9.4.1.1.3 Participants

Participants were 26 members of the general community (12 females, 14 males; $M = 30.0$ years, $SD = 22.3$). It was deemed unnecessary to use pilots, as the tasks under examination did not require any prior flying experience or familiarisation with a flight deck. Due to the positioning of the displays, participants were required to be

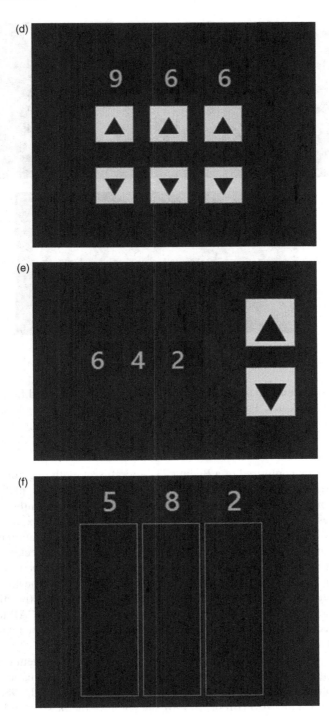

FIGURE 9.3 Continued

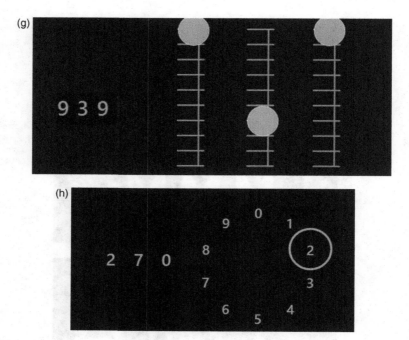

FIGURE 9.3   Continued

right-handed, between 160 and 190 cm, not wear bifocal lenses, and have no long-term chronic pain issues. Participants had an average right arm length of 77.1 cm ($SD = 6.8$) and height of 171.6 cm ($SD = 7.9$).

9.4.1.1.4   Equipment

A mock flight deck comprised of a Boeing 737 aircraft seat, with five-point harness, and three screens, was attached to a 6-axes motion simulation platform (see Figure 9.4). This is a 3 m × 2 m platform which can move independently along three translation axes (fore-and-aft, lateral, vertical) and three rotational axes (roll, pitch, yaw). This allowed for the simulation of turbulence and the configuration of screens akin to a current flight deck. Two 17.3" 16:9 LCD screens with projected capacitive (PCAP) touch were used for the panning tasks; one positioned directly in front of the participant, (centre; position of the Primary Flight Display) and the second to the right, parallel to the platform floor with a 15° rotation off horizontal (position of the Flight Director). A 16.3" 23:4 LCD glareshield screen also with PCAP touch was positioned directly above the centre screen and used exclusively for the number entry component (see Figure 9.5 for layout).

Objective evaluation methods were captured (e.g., time to completion and extra interactions with the touch screen). In addition, three subjective evaluation methods were also employed: workload via the NASA-TLX (Hart & Staveland, 1988; NASA, 1986), usability via the SUS (Brooke, 1996) and comfort via the Cornell University Questionnaire for Ergonomic Comfort (Hedge et al, 1999).

FIGURE 9.4 Physical layout of mock flight deck on simulator platform.

FIGURE 9.5   Layout of screens for the various tasks in mock flight deck.

Note. Pan tasks were performed on the lower centre and side screens. Number entry tasks were performed on upper centre screen.

9.4.1.2   Results

When considering objective performance of the pan task, the drag gesture was completed significantly faster and with less additional interactions than the two other gestures (tap, tap-and-hold). This was then supported by ratings for workload and usability, with drag tasks rated as requiring less workload and higher usability. While not an issue for light chop or light turbulence, there was however a considerable negative effect of moderate turbulence on the performance of the drag task when in the centre screen position. The results from the analysis suggest that the side screen position would be the most appropriate, and therefore recommended, location.

Objectively, tasks were completed faster and with less variability in performance in the side position than the front-centre position. It also produced lower discomfort ratings for participants, and this is supported by the ratings of workload and usability. This was especially evident in the performance of drag tasks, where it is thought that the angle of the screen in the centre (15° rotation back from perpendicular to the floor), made the drag manoeuvre more difficult as participants' wrists were completely extended. This was compared to a side position (screen was rotated 10° forward from parallel to the floor) which required users to lean over onto the screen so that participants' wrists and hands were kept in a position more closely resembling a neutral wrist. This makes it easier to perform the tasks and reduced physical workload. Similarly, in the tap-and-hold, users were required to maintain the extended wrist position for a longer period without release and this is evident in the workload and usability scores for those tasks and the associated significant interactions.

Overall, turbulence had a greater effect on number entry tasks, although when this was examined further, it was suggested that it is the specific gesture that was driving the effect. Single tap gestures, such as what was required for the keypad, buttons below, and buttons astride, were not affected by turbulence. Meanwhile, the more complex gestures of swipe, slide and drag (seen in the vertical swipe, vertical slide and clockface, respectively) were deemed to be inferior to the current dial in their time to completion, workload and usability. For this, they also appeared to have been more heavily influenced by the effects of turbulence, which was not seen in the pan task.

The task significantly impacted by turbulence was the vertical swipe task, which saw large increases in performance variability, the number of additional interactions required to complete the task, and the time taken to complete the task with increases in turbulence. This supports the findings of Coutts et al (2019) which found that the swipe action had lower accuracy than other gestures. This may be due to the difference between a swipe and drag gesture, where both are considered to be a single finger gesture, swiping is inherently a lighter, 'flick' and dragging implies constant contact with the screen to complete the gesture (Jeong & Liu, 2017). The faster the screen is swiped, the larger the change in the associated screen interface. The 'lighter' action of the swipe appears to be more susceptible to turbulence than the drag, where the constant contact with the screen may be a preventative factor against turbulence.

Second, Burnett et al (2013) suggest that the intended direction of the action influences the swiping duration and speed of the swipe. While this configuration had swiping occurring only in an up-down direction, in Burnett et al's study, down swipes were slower than upward swipes, which is associated with a smaller change in the screen. While not the same effect as turbulence, moving from a static to driving condition enhanced this gesture speed finding (Burnett et al, 2013). This could account for the differences in the performance across turbulence in the vertical swipe task.

9.4.1.2.1 Conclusions and recommendations

The single tap keypad design was the most supported new design and appears to outperform the current mechanical dial interface. The keypad design was the fastest task to be completed, and with limited additional interactions, suggesting fewer errors or inadvertent presses. Interestingly, the two remaining tap interfaces (arrows below and arrows aside) were completed at a similar speed (differences were not significant), and not significantly different from that of the keypad. Therefore, it would suggest that any tap-based integration would enhance performance over the current mechanical dial for number entry, regardless of turbulence. The workload and usability ratings are therefore important when considering which of these tasks would be best integrated into the flight deck. The keypad task was reported to have a significantly lower workload and significantly higher usability than the mechanical dial, while buttons astride and below were not significant from the dial. The results of the keypad assessment also suggest there would be less variability in individual users. This would reduce the potential for error in the number entry task. Taking this into account, the keypad is the preferred integration format.

The research makes a secondary contribution through the study of the pan task, allowing for a direct comparison between panning gestures on the same tasks. Although not currently part of the flight deck design, this is a new potential feature which may be used as the flight deck moves toward becoming more interactive. For example, a pan function may be employed in adjusting map displays in navigational way point selection or selecting the relevant 'app' or instrumentation required in an integrated flight-deck display. Being a new integration that cannot be compared against an existing form of interaction, it contributes to the understanding of the impact of turbulence on touchscreen interactions. Whilst the results were not as predicted, some light can be shed on how this specific task is completed under turbulence. This highlights the need for further turbulence-based testing of new touchscreen tasks, to determine how each could potentially be affected.

The most important contribution of the current study is the difference found in the type of task and the optimum gesture to be completed. The combination of subjective and objective evaluations in the simulated environment provided the most comprehensive evaluation of the tasks that could be performed. A limitation exists in that participants were only using their dominant hand (right-handed participants on the right-hand screen), so performance, usability, workload and comfort may be impacted if the non-dominant hand is used (e.g., left-handed participants on the right-hand screen or right-handed participants on a left-hand screen).

This research further supports the call for a confirmation function when higher levels of turbulence are present (Coutts et al, 2019), which could be activated when high turbulence is encountered. This would ensure that any accidental presses (such as the increased additional screen interactions seen at higher turbulence) are not registered as the final selection, particularly for safety-critical processes, without pilot confirmation.

9.5 CONCLUSIONS

Each of the evaluation techniques outlined within this chapter was critical in the continued development of technological concepts as well as the identification of future research avenues. They also demonstrated how end-user input is valuable throughout the design process – not only within simulation studies, but also in the wire-frame evaluations that can occur much earlier on within the design process.

The wire-frame heuristic evaluation (Section 2) presented useful feedback from airline pilots on their observations and interpretations of the interfaces which was used to guide design selections and improvement. The use of the wire-frames made for a cost-effective analysis of the initial designs prior to full-scale software development.

The simulated study explored the use of a proposed indicator for oil starvation (Section 3) and identified the issues related to information overload and how this may impact the ability of pilots to make decisions in a safe environment. Thus, the use of a simulator reduced some of the risks associated with testing in a real aircraft, whilst also providing some degree of realism associated with the event. The use of pilots in this evaluation was pivotal, as such a task with non-pilot participants, who lacked the knowledge of the standard operating procedures, would have likely been inconclusive.

The second simulator evaluation (Section 4) highlights that the intention of the task to be completed plays a pivotal role in its design. The initial evaluation of touchscreen tasks out of context, did not provide sufficient information for performance evaluation. Without considering the specific task that was required to be performed, evaluation of the interface may have not identified the issues with the dragging task for number entry. This highlights the need to empirically test tasks in moving-base simulators when evaluating and designing human–machine interfaces, as a 'one-gesture-suits-all' approach does not apply.

A notable limitation to the end-user analysis exists in the sample size of some of the evaluations. By relying on a specific population of pilots, it was harder to recruit participants as compared to members of the general population. These sample sizes have dictated the analyses that were able to be performed, limiting the potential significance and impact of the results. However, the smaller scale, "proof of concept" trials, as discussed in Section 3, have allowed for an in-depth analysis of a subset of the pilot population and generated a number of lessons to be taken forward into future testing. Therefore, the small sample sizes should not be overlooked in terms of their contribution. Without them, incremental modifications, which form a central feature of the design process, would not be possible.

This chapter has presented the different options for design evaluation, the outputs of which may require further redesign and modelling, following the iterative nature of the design process noted in Chapter 5. In further iterations the use of further user input is recommended using the methods and measures previously mentioned within this book. The next chapter will conclude this book by bringing together the work presented throughout the previous chapters of the book and provides some concluding remarks.

10 Conclusion

10.1 INTRODUCTION

This concluding chapter brings together the work presented across the chapters of this book and presents some of the key outputs and contributions. Throughout this book, the reasoning, methods and processes involved in developing new flight-deck technologies from a user-centred approach have been described. We highlighted that there are three key stages involved in this approach: designing, modelling and evaluating. The inclusion of the end-user throughout each of the three stages is imperative to the successful design of new technologies, especially on the flight deck. Within this chapter we summarise each of the chapters in turn and provide a high-level synopsis of the important elements of user-centred design on the flight deck. Recommendations for applying and replicating this process are then presented.

10.2 BOOK SUMMARY

The first chapter of this book provided an introduction to the subject area and the importance of applying Human Factors principles to the flight deck. This included a high-level summary of the background to this work and what the main contributions of the book will focus on. The second chapter continued to provide more background information relating to the principles and processes involved in Human Factors best practice. It provided an overview of how to design and conduct experimental research with application to the aviation domain. Importantly, the distinctions between different types of research were discussed here, including the distinction between qualitative and quantitative research and the varying applications of these two approaches. The user-centred design process is also introduced within this chapter, setting the scene for the following chapters of the book which develop the user-centred design process for future flight-deck technologies. This second chapter is particularly useful to consider at the very early stages of the research process and the principles discussed within it should be understood by all involved in the design process.

The third chapter presented the methods and measures that are relevant to the study of Human Factors in the aviation domain. A number of different measures that can capture the end-user's mental state, performance and functionality were detailed. These should be carefully considered when determining what type of research is required and what the research outputs intend to inform. For example, is the aim to determine the ability of the user to perform certain tasks without error?, or are more subjective measures of attention, fatigue or trust of importance? The second part of this chapter presented a range of different methods that enable user-centred design to be undertaken. These methods were described here and then applied throughout the later chapters of the book. Consideration for the different aspects of human performance and the methods used to study them should be undertaken early on within a project to help shape the research questions. This early consideration should also take account of the ways in which usability is understood, as outlined in Chapter 4. Chapter 4 presented a set of usability criteria that are specific to the aviation domain and the design of aircraft cockpits. The literature presented, and the criteria listed, provide a valuable foundation for understanding what it means to design and develop usable technologies for the flight deck.

Chapter 5 presented an overview of the user-centred design process that is then described in more detail throughout the remaining chapters of this book. It included a useful diagram of the process, highlighting the stages involved, the methods and the actors that should be involved across each of the stages. This is a useful reference to guide the process of designing new aviation technologies. Each of the stages are then focused upon in later chapters, starting with the design requirements stage in Chapter 6. This chapter presented the two ways in which design requirements may be generated with examples to explain the process. The first method is for advancing current systems, where the functions of the system are re-designed. The example used here was the re-design of the Flight Management System (FMS). A hierarchical task analysis (HTA) was first applied to map out all of the tasks involved in the case study example of entering flight navigation information into the FMS. The identification of all tasks enabled the Systemic Human Error Reduction and Prediction Approach (SHERPA) to be applied to review where there was opportunity for error. These error points were then the focus of the requirements for the new system. The second approach for generating requirements lends itself to when a new avionic system is required, wherein no previous system can be used as a starting point. Here an interview method with pilots is presented with the case study of an oil starvation avoidance alert. This qualitative method generated in-depth information from pilots that was used to inform design requirements. The Perceptual Cycle Model was applied here to outline the interaction between the user (pilot) and their environment in relation to decision-making when facing an engine oil leak. Such theoretical applications can help structure qualitative analyses and identify where new technology can help to support the pilot.

Chapter 7 presented the Design with Intent method which involves the user in the design of novel and unconstrained new designs. Through design cards, users are prompted to think about the design of new technologies and to draw these out as rudimentary designs. The case study from the previous chapter of the engine oil leak detection system was provided to show how the designs can be generated through workshops with users (in this case commercial pilots). Example images of the drawings from these workshops showed the standard that is expected from these workshops. This is a highly engaging method that can provide rich user input when conducted effectively.

The second part of this chapter then compared the designs to the requirements that were generated using the SHERPA method, as presented in Chapter 6. This comparison is important to identify if the designs align with the requirements. Importantly, it was identified that the requirements outlined by researchers and Human Factors and aviation professionals are not always identical to those of the end-user. A comparison between the end-users' designs and the Human Factors requirements shows some divergence in some areas. For further detail on this please see Parnell et al (2021a).

Chapter 8 presented the modelling elements of the design process using two approaches: engineering integration modelling and human decision modelling. Both approaches are important, but their application depends on the design and context of use. The engineering integration modelling approach presented the utility of Operator Event Sequence Diagrams (OESDs) to model the interaction between all actors involved in the system under development and how they enable each of the required tasks to be conducted. This is important as it considers the wider context of use and the highly interactional nature of the cockpit, its operators and the technology. The FMS case study was returned to, which highlighted how a future system would integrate with other technologies on the flight deck and the tasks that the user must perform. Comparisons between current and future systems within the OESDs showed how actions and interactions will need to change within the wider flight-deck system. The second part of Chapter 8 presented the decision modelling that is required when determining how new technologies may influence the decisions that users make. This was discussed in relation to the design of a novel decision aid that provides information to the pilot on engine status following fan blade damage. Three different decision models were presented: the Recognition Primed Decision Model (RPDM; Klein, 1989), Decision Ladders (Rasmussen, 1983) and the Perceptual Cycle Model (PCM; Neisser, 1976). Each model was applied to the same case study to demonstrate the differences in how users' decisions can be understood and the input they can have to the design process. A comparison of all models highlighted the particular utility of the PCM in capturing both user and system performance. More detail on this can be found in Parnell et al (2022a).

The final stage, evaluation, was presented in Chapter 9, which included the different ways that designs can be evaluated depending on the fidelity of the design and its intended context of use. The use of the heuristic evaluation method was detailed in relation to the design of a cool-down/warm-up indicator which could provide pilots with more accurate information on the temperature of the engines before take-off and after landing. The heuristic evaluation was useful here as the designs were at the wire-frame stage and this method is cheap and relatively straight forward to undertake, without the need for more developed prototypes. The method can also be applied to review multiple design options in relation to the usability heuristics to make comparisons and identify which design aspects should be taken forward. The next stage for these designs is then to test them within simulated environments. This requires more advanced prototypes and integration into the cockpit. A case study example of a flight simulator was presented in Chapter 9 with the oil starvation alert that was detailed throughout previous chapters of the book. The design of the simulator study utilised the experimental considerations, outlined in Chapters 2 and

3, and pairs of qualified airline pilots were used to assess their interactions within the simulated environment. This evaluation technique is more complex and time intensive than the heuristic evaluation, but well-designed studies can have a significant impact on the development of new technological designs. The final part of this chapter presented the evaluation process in respect to the context of use. Within the aviation domain it is vitally important to consider the wider system of systems that comprise the domain when doing evaluative work. When interacting with cockpit technologies the presence of turbulence is a commonplace environmental factor that must be included within the evaluation of any new design. When we consider the utility of touchscreen technologies within the flight deck their use under differing turbulent conditions needs to be carefully considered. Here, we presented a number of different ways in which touchscreens can be interacted with and evaluated them under different levels of turbulence. This is a first step towards determining the feasibility of future cockpit designs and what measures may need to be put in place to ensure they are safe to use across the varying contextual elements that comprise the aviation domain.

10.3   REFLECTIONS ON USER-CENTRED DESIGN ON THE FLIGHT DECK

This book has presented a number of different case study examples to demonstrate the different stages of the user-centred design process and the methods that comprise it. The case studies were developed in collaboration with industrial partners from aircraft and aerospace manufacturers and test pilots. Through the development of this design process, we have received input, support and guidance from a number of different stakeholders, including aircraft engineers, commercial airline pilots (of varying experience and age), test pilots and Human Factors practitioners. The outputs presented within this book have therefore undergone the scrutiny and support of numerous different parties involved in the design of future flight-deck technologies.

Notably, the work has been well received by many of the user groups that we have worked with. The aircraft and aerospace manufacturers were very receptive to the methods and measures that we proposed and were eager to understand how they can align within their own design processes while also including the end-user aspect that they had previously not well defined or understood. Indeed, this has been a motivation for writing this book, in the hope that it will provide a practical guide for industry colleagues to embed user-centred design at the heart of what they do. The airline pilots we recruited and those we worked with were generally very encouraging and receptive to the research, and they valued being involved within the design process. Similarly, the test pilots were encouraging of the tools and helped develop the thoughts and ideas of the commercial pilots, while also providing more in-depth knowledge of the functions of the flight deck. Central to this is the types of methods used and the involvement of the different user groups at key stages in the design process. Some of the key lessons learnt over the course of the programme of work presented in this book are discussed below.

10.3.1 Methodological Lessons

10.3.1.1 Qualitative Research

Qualitative data collection and analysis can give a rich source of end-user data that can be valuable to the design process. Within this book, we highlighted the utility of user interviews at the design requirements stage of the design process in Chapter 6. The Schema World Action Research Method (SWARM; Plant & Stanton, 2016) has been used extensively throughout this work to structure the interviews with pilots. The method has a pre-defined list of interview prompts that are specifically related to the aviation domain, and were in fact designed in a user-centred way, based on the decision-making processes of pilots (Plant & Stanton, 2016). A number of publications have been written based on this method to capture pilot decision-making and generate design requirements (Banks et al, 2020a; Parnell et al, 2021b, 2022a). The real benefit of this approach is that is maps directly onto the Perceptual Cycle Model (Neisser, 1976) and therefore allows for data to be coded and used to populate the model. This structured approach to qualitative analysis is highly beneficial to ensure the rich data source is well constructed and understood. We found that the high level of training and knowledge that pilots hold led to a lot of similarities in their responses to the interview prompts and therefore we were able to map out the perceptual cycle with relative ease. Other research has started to apply the method and model to other domains such as automotive automation (Banks et al, 2018a) and unmanned aerial vehicles (Parnell et al, 2022b), which has required adaptations to account for the higher levels of variability in these user groups.

Qualitative assessments were also used within the evaluation stage of the design process to complement some of the quantitative measures. Within the heuristic evaluation of the wire-frames, qualitative reports were used to support and explain the numeric ratings that were given to each of the designs. In the flight simulator study the pilots were interviewed after each of the flight trials to capture their feedback and decision-making processes throughout the flight. The richness of this qualitative data was used to inform the quantitative data analysis in both cases. This led to impactful results and was useful to the later stages of the design process.

Often qualitative research may be overlooked within the engineering discipline, or not conducted rigorously, but when taking a user-centred approach it is vital to include qualitative data, ideally in combination with performance metrics, to build a broader picture of the user experience. The work presented within this book has provided evidence for this, as well as methods that can be used to capture and analyse qualitative date within the aviation domain.

10.3.1.2 Experimental Setting

A number of different experimental methods have been detailed within this book. The different experimental approaches and settings require different considerations, and it is important that these are recognised from the outset. The generation of design requirements in Chapter 6 utilised interviews to capture qualitative insights from users, as detailed in the previous section. When using this method, a clear understanding of the intended purpose and intentions of the interviews (i.e. research question) needs to be made clear from the outset. We utilised scenarios/use cases to provide a context for the interviews and enable specific details to be captured. A scenario that the participants will be familiar with will enable more rich details to be collected. Semi-structured questions such as those provided by SWARM (Plant & Stanton, 2016) are useful as they give a structure to the interview whilst also allowing some adaption to unanticipated and interesting information that may arise. The setting for interview data collection should be one where the interviewee feels confident to discuss the topic of interest. To enable more participants to take part, we travelled to airports where the pilots could be interviewed with ease. The COVID-19 pandemic has also meant that online platforms have been a lot more common place and we found that the interview was not compromised when conducted online. The use of PowerPoint presentations on shared screens can enable all parties to easily follow the interview questions. Online platforms that record and even transcribe the interviews are also very beneficial to this process. Online settings can also increase participant uptake and allow for a more inclusive approach to data collection, with more balanced samples than perhaps in-person data collection can allow.

Design workshops, such as those discussed in Chapter 7, can also be run using online platforms where needed. However, in our experience it is more beneficial to run these in person where possible as they can allow for more creativity and inspiration, particularly when users are working in collaboration with their designs. In our experience, pen and paper inspired more creativity than online forums, so we recommend identifying locations and times of day when end-users can come together to participate.

The simulator testing conducted in Chapter 9 is a complex experimental setting that requires a lot of organisation, planning and trialling to ensure effective outputs. The preparation for the experiments detailed in the case study in Chapter 9 required significant planning and a clear understanding of the testing requirements by the Human Factors practitioners, aerospace manufacturer, simulator technicians and the participants. The number of trials that could be run per participant was a significant issue as multiple designs and environments were of interest, but time considerations meant that only a limited number could be run before the participants would struggle to stay engaged. The training condition was imperative as the participants had never flown the flight simulator before and it was quite different to the simulators that they normally fly as part of their training, with new touchscreen interfaces and novel designs. The familiarisation time in an experiment such as this can be quite long and therefore this should be also considered when designing experiments. Including a baseline condition is also vitally important as this allowed us to compare the effect of the experimental conditions. There were a number of technical faults that occurred throughout the testing process and this is often to be expected. It is important that the correct people are on hand to fix the problems and prevent time wastage. However, it also important to explain the experimental nature of the work to the participants to set their expectations.

10.3.1.3   Human Participants

Human participation is central to user-centred design, and the characteristics of the user need to be carefully considered when recruiting and experimenting with human participants. Within the work presented in this book we sought commercial airline pilots as our end-user. We also used test pilots, however, this was as a secondary data source after commercial pilots were used, as test pilots have more knowledge and experience than standard pilots and the technologies we were designing were to be used by commercial pilots. The time constraints on pilots meant that accessing them was challenging. The recruitment process targeted pilot forums and groups that commercial pilots are members of. As mentioned in the previous section, we travelled to airports to enable pilots to easily take part without them having to travel too far out of their way. The exception to this was the flight simulator testing, as this was not possible. Here, we left the testing window open for a long period to give flexibility to the participants and always offered a monetary incentive as a token of gratitude and to cover any incurred expenses.

It is also important to consider the demographics of the user sample. We found that older retired pilots were the most likely demographic to volunteer their time to participate as they had more free time. However, when reviewing new aviation technologies this demographic is not fully representative of the end-users of the future (i.e. younger pilots, who have grown up with technology in all aspects of their lives). Furthermore, we had difficulties in recruiting female participants due to the male dominance within the aviation domain. However, it is vitally important that equal, or at least representative, gender samples are captured within the user-centred design process to avoid a gender-data gap whereby the influence of gender is not realised in the research process (Madeira-Revell et al, 2021). The default male bias that is evident across transportation domains is very pertinent within the aviation domain due to the historic absence of females involved in the design or flight of aircraft (Parnell et al, 2022c). Yet, by including more females in the design process this can start to be undone and more females can be encouraged into the domain. Within this work we sought out female participants through social media and targeted advertisements and found that females were generally very happy to be approached. Considering the time of testing and location is also important to ensure that recruitment is inclusive and open to all possible users.

10.4   RECOMMENDATIONS

Below we provide a summary of the key recommendations made across the three main aspects of the user-centred development process presented in this book: designing (including generating requirements), modelling and evaluating.

10.4.1   DESIGN

- When considering designing new systems always define design requirements from the outset.
- There are two main ways in which design requirements can be generated, depending on the type of design that is desired: advancing current system or defining new concepts.
- When advancing new systems, the current system can be used as a starting point to generate requirements through task modelling and failure analysis.
- When defining new concepts, the users can be consulted to identify the requirement for the new technology and how it would integrate into their current understanding and use of the system.

- Design requirements must account for the context of use across a variety of scenarios, within the aviation domain this includes the other technologies on the flight deck and the impact of turbulence on physical interactions.
- Design requirements must include input from the users and how they use the system. Qualitative methods such as interviews are a valuable tool here and presenting scenarios can help to define certain concepts of operation.
- When designing with end-users the use of workshops in person is encouraged. These workshops should encourage out-of-the-box thinking at the early stage of the design process. The Design with Intent method is particularly useful here.
- Collaboration between pilots and aerospace manufacturers and/or engineers is a powerful way to ensure end users' needs are represented within the design. Grouping them together in design workshops can facilitate conversations on user needs at an early stage of the design process.
- Pilot end-users and Human Factors practitioners come from different perspectives when defining design requirements and generating designs. Therefore, reviewing and comparing the outputs, as was done in the second part of Chapter 7, is imperative. Conflicting requirements need to be tackled at this early stage in the process to limit costly changes later on.
- As many different stakeholders as possible should be involved in the early stages of the design process, through the methods presented in Chapters 6 and 7 we set out ways in which to facilitate this.

10.4.2 MODEL

- Modelling how the new design and/or technology will integrate within a current or future system is a vital step in the design process. This can be done in a low-cost way and should also incorporate feedback from the user, manufacturer and Human Factors practitioners to review possible issues early on.
- Modelling the interactions between different actors and objects within the system can be done through OESDs. This enables the engineering integration to be modelled in a visual and easy-to-review way. This was particularly valued by systems engineers to review the wider system of the technologies integration.
- OESDs of current and future systems can be conducted to review the impact of the technology and how it will change the interactions between different actors. This is particularly useful when automation is introduced.
- Modelling of the users' interactions with the technology should also be conducted at a psychological, or human behaviour, level. This can utilise decision-making and information-processing models where technology aims to provide new or different information to the user.
- There are multiple different psychological models that present different ways in which to model human behaviour. Reviewing each with respect to the aims of the work is encouraged, with an understanding of their limitations.
- In Chapter 8, we review three different decision models which each capture decisions in different ways, and this impacts the way in which they make design recommendations. However, it is important that the systemic impact of decision-making is realised and the PCM was best able to capture this.

- The PCM is a useful method for modelling and visualising human behaviour in relation to the information they receive in the world and how they act upon it. We therefore recommend it as a useful tool in modelling user interactions with new technological designs.

10.4.3 EVALUATE

- Evaluations should be conducted across the different levels of fidelity in the design development and there are different methodologies for these different stages of the design.
- Low-cost and time-saving evaluation methods should be applied at the wireframe mock-up stage, when designs are still in the conception phase. Heuristic evaluations should be used here and can be applied to compare multiple different designs without the need for full mock-ups.
- Nielsen's (1994) usability heuristics are a useful tool, the heuristics are standardised and have been applied with effect across multiple domains, including aviation.
- Different actor groups should engage in the heuristic evaluation; a combination of Human Factors experts and users can highlight different perspectives and areas for improvement. In the aviation domain users should include commercial airline pilots, not just test pilots.
- After heuristic evaluations are complete an improved design can be developed for higher fidelity testing, this can be integrated into a simulated environment to replicate the context of use. Within the cockpit, the interaction with other features on the flight deck is important to consider, so these should also be added to the simulator environment where possible.
- When designing simulator experiments with human participants it is important to consider the principles of research design presented in Chapters 2 and 3. Well-designed experiments are best placed to make meaningful recommendations and prevent wasting participants' time. Time spent getting the study design correct is invaluable.
- A representative sample is required for user evaluations, importantly this should include an equal gender split and data should be disaggregated by gender to assess any gender impacts. This will help to close the gender-data gap and encourage more females into the aviation domain.
- Quantitative metrics from simulator studies should be complemented with qualitative data from the users to capture their experiences of interacting with the designs. Using both qualitative and quantitative data in combination is a powerful tool to suggest where updates are required for successful design.
- Context of use is essential to determining the usability of a technology, as outlined in Chapter 4. Within the aviation domain, the flight deck is often subject to turbulent conditions which will impact on how usable an interface is on the flight deck. Ensure that this is considered early on in the design process.
- Chapter 8 provides data on a number of different ways of interacting with touchscreen technologies which are differentially affected by turbulence conditions. These interaction methods should be carefully considered within the design process.

- The design process is iterative and therefore once a design has been evaluated it can be reviewed against the design constraints again and the process can, and often should, be repeated to fine tune the design.
- The same or different users can be reused throughout the design process, depending on the aims of the process and what is under review. However, usually the best approach is to include as many diverse members of the user population as possible to ensure that designs are inclusive.

10.5   APPLICATIONS TO OTHER DOMAINS

This book has detailed a user-centred design process that is focused on the aviation domain, with case studies that directly relate to the design and development of future flight-deck technologies. However, Human Factors is a cross-disciplinary approach and therefore the methods and measures that have been applied throughout this book can be applied to other domains, and indeed they have been. Many of the methods, such as Design with Intent and the PCM have been applied to the automotive industry (Allison & Stanton, 2020; Banks & Stanton, 2018). The design of interfaces to provide enhanced information to the user is essential in many safety-critical domains such as healthcare, maritime, automotive, rail and the nuclear industry. The key difference across domains, however, is the user base. When conducting user-centred design, it is critical that the user base is well represented within the design process, including all types of users and their range of experiences and characteristics. Time and funding must be spent understanding the user population and ensuring that a representative sample has been recruited through a range of different recruitment methods.

10.6   CLOSING REMARKS

This book has presented an overview of the user-centred design process for designing, modelling and evaluating new technologies on the flight deck of future aircraft cockpits. Taking a number of different case study examples, the process for apply and analysing user-centred methods has been demonstrated. These case studies present some real-world challenges facing the future design of aircraft cockpits. The work presented within this book has been conducted in collaboration with aerospace manufacturers, aerospace engineers, commercial airline pilots, test pilots and Human Factors practitioners. Input from a large range of stakeholders is needed to set out the requirements of the design of new aviation technologies and to develop and integrate these designs. Yet, it is the end users' perspective which must be carefully considered throughout all stages of the design process. Often this can be forgotten when the demands of industry, time and cost are prioritised. This book aims to provide the tools and methods through which the end-user, in this case commercial airline pilots, can be fully integrated within the design process to ensure it is indeed user centred. We hope that this book is of use to all stakeholders in providing a Human Factors perspective on how to enhance the usability, and ultimately, safety of future avionic systems.

References

AAIB. (1990). Report on the accident to Boeing 737-400G-OBME near Kegworth, Leicestershire on 8th January 1989. Aircraft Accident Report 4/90. Department for Transport, HMSO, London.

Alexander, A. L., Kaber, D. B., Kim, S.-H., Stelzer, E. M., Kaufmann, K., & Prinzel, L. J. (2012). Measurement and modeling of display clutter in advanced flight deck technologies. *The International Journal of Aviation Psychology, 22*(4), 299–318

Allison, C.K., & Stanton, N.A. (2020) Ideation using the "Design with Intent" toolkit: A case study applying a design toolkit to support creativity in developing vehicle interfaces for fuel-efficient driving. *Applied Ergonomics, 84*:103026

Annett, J., Duncan, K. D., Stammers, R. B. & Gray, M. J. (1971). *Task Analysis. Department of Employment Training Information Paper 6*. London: HMSO.

Asmayawati, S., & Nixon, J. (2020). Modelling and supporting flight crew decision-making during aircraft engine malfunctions: developing design recommendations from cognitive work analysis. *Applied Ergonomics, 82*, 102953.

Avsar, H., Fischer, J. E., & Rodden, T. (2016). *Mixed method approach in designing flight decks with touch screens: A framework*. In 2016 IEEE/AIAA 35th Digital Avionics Systems Conference (DASC) (pp. 1–10). IEEE.

Australian Transport Safety Bureau. (2012). Engine oil leaks VH-OQG and VH-OQC en route Singapore to London, United Kingdom, 24 February and 3 November 2011. Retrieved from www.atsb.gov.au/media/3545749/ao-2011-034_final.pdf. Accessed 08/08/2018.

Australian Transport Safety Bureau. (2017). Engine malfunction and in-flight shutdown involving Boeing 777 A6-EGA. Retrieved from www.atsb.gov.au/media/5772230/ao-2016-113-final.pdf. Accessed 08/08/2018.

Avery, D. (2011). The evolution of flight management systems. *IEEE Software, 28*(1), 11–13.

Baber, C. & Stanton, N.A. (1994) Task analysis for error identification. *Ergonomics, 37*, 1923–1942.

Baber, C., & Stanton, N. A. (1996). Human error identification techniques applied to public technology: predictions compared with observed use. *Applied Ergonomics, 27*(2), 119–131.

Banfield, R., Lombardo, C. T., & Wax, T. (2015). *Design sprint: A practical guidebook for building great digital products*. "O'Reilly Media, Inc.".

Banks, V. A., & Stanton, N. A. (2018). Automobile automation: Distributed cognition on the road. CRC Press.

Banks, V. A., Stanton, N. A., & Harvey, C. (2014). Sub-systems on the road to vehicle automation: Hands and feet free but not 'mind' free driving. *Safety science, 62*, 505–514.

Banks, V. A., Plant, K. L., & Stanton, N. A. (2018a). Driver error or designer error: Using the Perceptual Cycle Model to explore the circumstances surrounding the fatal Tesla crash on 7th May 2016. *Safety Science, 108*, 278–285.

Banks, V. A., Plant, K. L., & Stanton, N. A. (2018b). Development of a usability evaluation framework for the flight deck. In Contemporary Ergonomics and Human Factors. *Proceedings of the International Ergonomics and Human Factors Conference*.

Banks, V., Parnell, K., Plant, K., & Stanton, N. (2019). Design with intent on the flight deck. In Contemporary Ergonomics and Human Factors. *Proceedings of the International Ergonomics and Human Factors Conference*.

Banks, V.A., Allison, C.K., Plant, K.L., Parnell, K.J., & Stanton, N.A. (2020a). Using SWARM to generate design requirements for new systems on the Open Flight Deck. Human

Factors and Ergonomics in Manufacturing & Service Industries. Online Version https://doi.org/10.1002/hfm.20869

Bartlett, F. C. (1932). *Remembering: A study in experimental and social psychology*. Cambridge University Press.

Bender, G. T. (1999). Touch screen performance as a function of the duration of auditory feedback and target size. PhD dissertation., Wichita State University.

Billings, C. E. (1991). Toward a human-centered aircraft automation philosophy. *The International Journal of Aviation Psychology, 1*(4), 261–270.

Billings, C. (1997). Aviation automation: The search for a human-centered approach. New Jersey: Erlbaum

Boeing. (2018, August 15). Cleared for takeoff ... Available at: www.boeing.com/features/2018/08/route-sync-08–18.page [Accessed 25 Sept 2018].

Brooke, J. (1996). SUS: A 'quick and dirty' usability scale. In P. W. Jordan, B. Thomas, B. A. Weerdmeester, & I. L. McClelland (Eds) Usability Evaluation in Industry, pp. 189–194. London: Taylor and Francis.

Brooks, F. A. (1960). Operational sequence diagrams. *IRE Transactions on Human Factors in Electronics, 1*, 33–34.

Burnett, G., Crundall, E., Large, D., Lawson, G., & Skrypchuk, L. (2013). A study of unidirectional swipe gestures on in-vehicle touch screens. In *Proceedings of the 5th International Conference on Automotive User Interfaces and Interactive Vehicular Applications, AutomotiveUI '13*. New York, USA, 22–29. https://doi.org/10.1145/2516540.2516545

Chapanis, A., (1999). *The Chapanis Chronicles*. Aegean Publishing Company: Santa Barbara, California.

Chatzoglou, P. D., & Macaulay, L. A. (1996). Requirements capture and analysis: A survey of current practice. *Requirements Engineering, 1*(2), 75–87.

Corno, F., Guercio, E., De Russis, L., & Gargiulo, E. (2015). Designing for user confidence in intelligent environments. *Journal of Reliable Intelligent Environments, 1*(1), 11–21.

Courteney, H. (1998). Assessing error tolerance in flight management systems. CAA (UK) Safety Regulation Group.

Coutts, L. V., Plant, K., Smith, M., Bolton, L., Parnell, K. J., Arnold, J., & Stanton, N. A. (2019). Future technology on the flight deck: Assessing the use of touchscreens in vibration environments. *Ergonomics, 62*(2), 286–304.

Daalhuizen, J. (2014). Method Usage in Design: How Methods Function as Mental Tools for Designers. Delft: Delft University of Technology.

Damodaran, L. (1996). User involvement in the systems design process – a practical guide for users. *Behaviour & Information Technology, 15*(6), 363–377.

Dekker, S. (2011). *Drift into Failure: From Hunting Broken Components to Understanding Complex Systems.* Surrey: Ashgate.

Dekker, S. (2014). *The field guide to understanding 'human error'.* Ashgate Publishing Ltd.

Denzin, N. K., & Lincoln, Y. S. (Eds.). (2011). *The Sage handbook of qualitative research.* Sage.

Dodd, S., Lancaster, J., Miranda, A., Grothe, S., DeMers, B., & Rogers, B. (2014). Touch screens on the flight deck: The impact of touch target size, spacing, touch technology and turbulence on pilot performance. *Proceedings of the Human Factors and Ergonomics Society 58th Annual Meeting*, 6–10.

Dorneich, M. C., Dudley, R., Letsu-Dake, E., Rogers, W., Whitlow, S. D., Dillard, M. C., & Nelson, E. (2017). Interaction of automation visibility and information quality in flight deck information automation. *IEEE Transactions on Human-Machine Systems, 47*(6), 915–926.

Eckert, C., & Stacey, M. (2000). Sources of inspiration: a language of design. *Design Studies, 21*(5), 523–538.

Eldredge, D., Mangold, S., & Dodd, R. S. (1992). *A Review and Discussion of Flight Management System Incidents Reported to the Aviation Safety Reporting System* (Final report no. DOT/FAA/RD-92/2). Washington, DC: Battelle/U.S. Department of Transportation, 1992.

Embrey, D. E. (1986). SHERPA: A systematic human error reduction and prediction approach. In *Proceedings of the international topical meeting on advances in human factors in nuclear power systems*. Knoxville, TE.

European Aviation Safety Agency. (2017, May 12) Certification Specifications and Acceptable Means of Compliance for Large Aeroplanes CS-25. Amendment 19.

FAA (Federal Aviation Authority). (2011). Advisory Circular 20–175 (AC 20–175). Controls for Flight Deck Systems. Aviation Safety. Washington, DC: Aircraft Certification Service, Aircraft Engineering Division.

Fastrez, P. & Haué, J-B. (2008). Editorial: designing and evaluating driver support systems with the user in mind. *International Journal of Human-computer Studies, 66*(3), 125–131.

Fénix, J., Sagot, J. C., Valot, C., & Gomes, S. (2008). Operator centred design: example of a new driver aid system in the field of rail transport. *Cognition, Technology & Work, 10*(1), 53–60.

Gander, P. H., Signal, T. L., Berg, M. J. V. D., Mulrine, H. M., Jay, S. M., & Mangie, J. (2013). In-flight sleep, pilot fatigue and psychomotor vigilance task performance on ultra-long range versus long range flights. *Journal of Sleep Research, 22*(2), 697–706.

Gander, P. H., Mulrine, H. M., van den Berg, M. J., Smith, A. A. T., Signal, T. L., Wu, L. J., & Belenky, G. (2015). Effects of sleep/wake history and circadian phase on proposed pilot fatigue safety performance indicators. *Journal of Sleep Research, 24*(1), 110–119.

Gawron, V. J. (2016). Overview of self-reported measures of fatigue. *The International Journal of Aviation Psychology, 26*(3–4), 120–131.

GE Aviation. (2018, May 2). Connected Flight Management System: A powerful solution in every phase of flight [Video]. YouTube. https://youtu.be/DYBCt-If4i4 [Accessed 25 Sept 2018].

Gore, J., Ward, P., Conway, G.E., Ormerod, T. C., Wong, B. L. W. & Stanton, N. A. (2018) Naturalistic decision making: navigating uncertainty in complex sociotechnical work. *Cognition, Technology and Work, 20*, 521–527.

Gould, J. D., & Lewis, C. (1985). Designing for usability: key principles and what designers think. *Communications of the ACM, 28*(3), 300–311.

Grady, M. P. (1998). *Qualitative and action research: A practitioner handbook*. Phi Delta Kappa International. Indiana, USA.

Graham-Rowe, D. (2011). Vibrating touchscreens that give you a push back. New Scientist, 211(2821), 20. www.newscientist.com/article/mg21128216-000-vibratingtouchscreens-that-give-you-a-push-back/

Grether, W. F., (1949). Instrument reading. 1. The design of long-scale indicators for speed and accuracy of quantitative readings. *Journal of Applied Psychology*, 33, 363–372.

Grote, G., Weyer, J. and Stanton, N. A., (2014). Beyond human-centred automation – concepts for human–machine interaction in multi-layered networks. *Ergonomics, 57*(3), 289–294.

Hancock, P. A. (2013). In search of vigilance: the problem of iatrogenically created psychological phenomena. *American Psychologist, 68*(2), 97.

Happich, J. (2015). Force sensing touchscreen leverages projected capacitive technology. www.eenewseurope.com/news/force-sensing-touchscreenleverages-projected-capacitive-technology

Harris, D. (2007). A human-centred design agenda for the development of single crew operated commercial aircraft. *Aircraft Engineering and Aerospace Technology, 79*(5), 518–526.

Harris, D. (2011). *Human performance on the flight deck*. Ashgate Publishing, Ltd.

Harris, D. (2016). *Human performance on the flight deck*. CRC Press.

Harris, D., & Li, W. C. (2011). An extension of the human factors analysis and classification system for use in open systems. *Theoretical Issues in Ergonomics Science, 12*(2), 108–128.

Harris, D. & Stanton, N. A. (2010) Aviation as a system of systems. *Ergonomics, 53*, (2), 145–148.

Harris, D., Stanton, N. A., & Starr, A. (2015). Spot the difference: Operational event sequence diagrams as a formal method for work allocation in the development of single-pilot operations for commercial aircraft. *Ergonomics, 58*(11), 1773–1791.

Harris, D., Stanton, N. A., Marshall, A., Young, M. S., Demagalski, J., & Salmon, P. (2005). Using SHERPA to predict design-induced error on the flight deck. *Aerospace science and technology, 9*(6), 525–532.

Hart, S. G., & Staveland, L. E. (1988). Development of N 837 ASA-TLX (Task Load Index): Results of empirical and theoretical research. In P. A. Hancock & N. Meshkati (Eds) Human Mental Workload, pp. 239–250. Amsterdam: North Holland Press.

Harvey, C., & Stanton, N. A. (2013). *Usability evaluation for in-vehicle systems*. CRC Press.

Harvey, C., Stanton, N. A., Pickering, C. A., McDonald, M., & Zheng, P. (2011). To twist or poke? A method for identifying usability issues with the rotary controller and touch screen for control of in-vehicle information systems. *Ergonomics, 54*(7), 609–625.

Heaton, N. O. (1992). Defining usability. *Displays, 13*, 147–150.

Hedge, A., Morimoto, S. & McCrobie, D. (1999) Effects of keyboard tray geometry on upper body posture and comfort. *Ergonomics, 42*(10), 1333–1349.

Helmreich, R. L. (2000). On error management: Lessons from aviation. *BMJ (Clinical research ed.), 320*(7237), 781–785.

Hix, D. & Hartson, H.R. (1993). *Developing user interfaces: ensuring usability through product and process*. John Wiley

Hoffman, E. R., & Sheikh, I. H. (1994). Effect of varying target height in a Fitts' movement task. *Ergonomics, 36*(7), 1071–1088.

Hollnagel, E. (2014). *Safety-I and Safety-II: The past and Future of Safety Management*. Farnham, UK: Ashgate.

Honn, K. A., Satterfield, B. C., McCauley, P., Caldwell, J. L., & Van Dongen, H. P. A. (2016). Fatiguing effect of multiple takeoffs and landings in regional airline operations. *Accident Analysis & Prevention, 86*, 199–208.

Hu, Y., Li, R., & Zhang, Y. (2018). Predicting pilot behavior during midair encounters using recognition primed decision model. *Information Sciences, 422*, 377–395.

Huddlestone, J., Sears, R., & Harris, D. (2017). The use of operational event sequence diagrams and work domain analysis techniques for the specification of the crewing configuration of a single-pilot commercial aircraft. *Cognition Technology & Work, 19*, 289–302.

Huddlestone, J. A., & Stanton, N. A. (2016). New graphical and text-based notations for representing task decomposition hierarchies: towards improving the usability of an Ergonomics method. *Theoretical Issues in Ergonomics Science, 17*(5–6), 588–606.

Hutchins, E. (1995). *Cognition in the wild*. MIT Press.

International Air Transport Association. (2015). FMS data entry error prevention best practices. Available at: www.iata.org/whatwedo/safety/runway-safety/Documents/FMS-Data-Entry-Error-Prevention-ed-1–2015.pdf. [Accessed 24 Sept 2018].

ISO (International Organization for Standardization). (1996). DD 235. Guide to in-vehicle information systems.

ISO. (1998) Ergonomic requirements for office work with visual display terminals (VDTs) — Part 11: Guidance on usability (ISO 9241–11). Retrieved from www.iso.org/standard/16883.html

ISO. (2019). Ergonomics of human-system interaction — Part 210: Human-centred design for interactive systems (ISO 9241–210). Retrieved from www.iso.org/standard/77520.html

Jenkins, D. P., Stanton, N. A., Salmon, P. M., Walker, G. H., & Rafferty, L. (2010). Using the decision-ladder to add a formative element to naturalistic decision-making research. *International Journal of Human–Computer Interaction, 26*(2–3), 132–146.

Jeong, H., & Liu, Y. (2017) Effects of touchscreen gesture's type and direction on finger-touch input performance and subjective ratings. *Ergonomics, 60*(1), 1528–1539.

Johnson, G. I. (1993). Spatial operational sequence diagrams in usability investigation. *Contemporary Ergonomics*, 444–450.

Jones, D. G. (2000). Subjective measures of situation awareness. In M. R. Endsley & D. J. Garland (Eds.), *Situation awareness analysis and measurement* (pp. 113–128). Lawrence Erlbaum Associates Publishers.

Jordan, P. W. (1998). Human factors for pleasure in product use. *Applied Ergonomics, 29*, 25–33

Kaber, D. B., Riley, J. M., & Tan, K. W. (2002). Improved usability of aviation automation through direct manipulation and graphical user interface design. *The International Journal of Aviation Psychology, 12*(2), 153–178.

Kaempf, G. L., Klein, G. (1994). Aeronautical decision making: The next generation. In Johnston, N., MacDonald, N., Fuller, R. (Eds.), Aviation psychology in practice (pp. 223–254). Aldershot, UK: Avebury.

Kirwan, B., & Ainsworth, L. K. (Eds.). (1992). *A guide to task analysis: the task analysis working group.* CRC press.

Klein, G. A. (1989). Recognition-primed decisions. In W. B. Rouse (Ed.), Advances in man-machine systems research (Vol. 5, pp. 47–92). Greenwich, CT: JAI Press.

Klein, G. (2000). Cognitive task analysis of teams. In *Cognitive task analysis* (pp. 431–444). Psychology Press.

Klein, G. A. (2008). Naturalistic decision making. *Human Factors, 50*(3), 456–460.

Klein, G. A., & Calderwood, R. (1996). Investigations of Naturalistic Decision Making and the Recognition-Primed Decision Model. Klein Associates Inc. Yellow springs, Ohio, USA.

Klein, G. A., Calderwood, R., & Macgregor, D. (1989). Critical decision method for eliciting knowledge. *IEEE Transactions on Systems, Man, and Cybernetics, 19*(3), 462–472.

Kujala, S. (2003). User involvement: a review of the benefits and challenges. *Behaviour & Information Technology, 22*(1), 1–16.

Kurke, M. I. (1961). Operational sequence diagrams in system design. *Human Factors, 3*(1), 66–73.

Landau, K. (2002). Usability criteria for intelligent driver assistance systems. *Theoretical Issues in Ergonomics Science, 3*(4): 330–345.

Lawton, R. & Ward, N. J. (2005). A systems analysis of the Ladbroke Grove rail crash. *Accident Analysis and Prevention, 37*, 235–244.

Lee, J. D., & See, K. A. (2004). Trust in automation: Designing for appropriate reliance. *Human Factors, 46*(1), 50–80.

Lintern, G. (2010). A comparison of the decision ladder and the recognition-primed decision model. *Journal of Cognitive Engineering and Decision Making, 4*(4), 304–327.

Lipshitz, R. (1993). Converging themes in the study of decision making in realistic settings. In G. A. Klein, J. Orasanu, R. Calderwood, & C. E. Zsambok (Eds.), *Decision making in action: Models and methods* (pp. 103–137). Norwood, NJ: Ablex.

Lipshitz, R. & Shaul S.B. (1997) Schemata and mental models in recognition-primed decision making In. Zsambok C.E, & Klein G. (Eds.), *Naturalistic Decision Making*, Lawrence Erlbaum, Mahwah, NJ, pp. 60–72

Lipshitz, R., Klein, G., Orasanu, J., & Salas, E. (2001). Taking stock of naturalistic decision making. *Journal of Behavioral Decision Making, 14*(5), 331–352.

Lockton, D. (2017). *Design with intent: Insights, methods and patterns for designing with people, behaviour and understanding*. Sebastopol, CA: O'Reilly.

Lockton, D., Harrison, D., & Stanton, N. A. (2010). The design with intent method: A design tool for influencing user behaviour. *Applied Ergonomics, 41*(3), 382–392.

Lockton, D., Harrison, D., & Stanton, N. A. (2013). Exploring design patterns for sustainable behaviour. *The Design Journal, 16*(4), 431–459.

Lockton, D., Harrison, D., Holley, T., & Stanton, N. A. (2009). Influencing interaction: Development of the design with intent method. *Proceedings of the 4th International Conference on Persuasive Technology*, p.5, ACM, 2009.

Lorenzo-del-Castillo, J. A., & Couture, N. (2016). The aircraft of the future: towards the tangible cockpit. In *Proceedings of the International Conference on Human-Computer Interaction in Aerospace, 11*, 1–8.

Madeira-Revell, K. M., Parnell, K. J., Richardson, J., Pope, K. A., Fay, D. T., Merriman, S. E., & Plant, K. L. (2021). How can we close the gender data gap in Transportation Research? *Ergonomics SA: Journal of the Ergonomics Society of South Africa, 32*(1), 19–26.

Marrenbach, J., Pauly, M., Kraiss, K. F. (1998). Design of a user interface for a future flight management system. *IFAC Proceedings, 31*(26), 305–309.

Marshall, A., Stanton, N. A., Young, M., Salmon, P. M., Harris, D., Demagalski, J., Waldmann, T., Dekker, S.W., (2003). *Development of the Human Error Template – a New Methodology for Assessing Design Induced Errors on Aircraft Flight Decks*. Final Report of the ERRORPRED Project E! Department of Trade and Industry, London.

Matlin, M. (2004). *Cognition* (5th Ed). Harcourt College Publishers.

Matthews, M. L., Bryant, D. J., Webb, R. D., & Harbluk, J. L. (2001). Model for situation awareness and driving: Application to analysis and research for intelligent transportation systems. *Transportation Research Record, 1779*(1), 26–32.

McLean, S., Read, G. J., Hulme, A., Dodd, K., Gorman, A. D., Solomon, C., & Salmon, P. M. (2019). Beyond the tip of the iceberg: using systems archetypes to understand common and recurring issues in sports coaching. *Frontiers in Sports and Active Living, 1*, 49.

McCormick, E. J., & Sanders, M. S. (1982). *Human factors in engineering and design*. McGraw-Hill Companies.

McCurdy, M., Connors, C., Pyrzak, G., Kanefsky, B., & Vera, A. (2006). Breaking the fidelity barrier. In *Proceedings of the SIGCHI Conference on Human Factors in Computing Systems – CHI '06* (pp. 1233–1242). New York: ACM

Medina, A. L., Lee, S. E., Wierwille, W. W., & Hanowski, R. J. (2004). Relationship between infrastructure, driver error, and critical incidents. In *Proceedings of the Human Factors and Ergonomics Society 48ᵗʰ Annual Meeting*, 2075–2080.

Mejdal, S., McCauley, M. E., & Beringer, D. B. (2001). Human factors design guidelines for multifunction displays. Available at: www.faa.gov. Accessed 23.08.2017

Mentour Pilot (2015, August 26). Initial FMC set-up – tutorial [Video]. YouTube. www.yout ube.com/watch?v=rmJsZqt6LP8 Accessed 25 Sept 2018

Mentour Pilot (2017, August 25). Full FMC set-up – Boeing 737NG [Video]. YouTube. www. youtube.com/watch?v=yN9NafwGOtk Accessed 25 Sept 2018

Mosier, K. L., & Manzey, D. (2019). Humans and Automated Decision Aids: A Match. In (eds) Mouloua M. & Hancock P.A., *Human Performance in Automated and Autonomous Systems: Current Theory and Methods*, CRC Press, Chicago, pp. 19–41.

Mosier, K. L., & Skitka, L. J. (1996). Human Decision Makers and Automated Decision Aids: Made for Each Other? In (Eds.), Parasuraman R. & Mouloua M. *Automation and Human Performance: Theory and applications*, New York, NY: CRC Press. pp. 120.

NASA (National Aeronautics and Space Administration). (1986). NASA Task Load Index (NASA-TLX), Version 1.0: Paper and Pencil Package. Moffett Field, CA: NASA-Ames Research Center, Aerospace Human Factors Research Division.

National Transportation Safety Board (2010). Loss of Thrust in Both Engines After Encountering a Flock of Birds and Subsequent Ditching on the Hudson River, US Airways Flight 1549 Airbus A320–214, N106US, Weehawken, New Jersey, January 15, 2009. Washington D.C. Online source: www.ntsb.gov/investigations/AccidentReports/Reports/AAR1003. pdf (Accessed 08/06/2020)

Naikar, N. (2010). A Comparison of the Decision Ladder Template and the Recognition-Primed Decision Model. DSTO-TR-2397. Air Operations Division. Defence Science and Technology Organisation. Australian Government Department of Defence. Victoria, Australia

Neisser, U. (1976). *Cognition and Reality.* W.H. Freeman and Company, San Francisco.

Nielsen, J. (1992). Finding usability problems through heuristic evaluation. *Proceedings of the SIGCHI Conference on Human Factors in Computing Systems*, Monterey, California June 1992, p. 373–380.

Nielsen, J. (1994). Heuristic evaluation. In Nielsen, J., and Mack, R.L. (Eds.), *Usability Inspection Methods*, John Wiley & Sons, New York, NY.

Neville, T. J., Salmon, P. M., & Read, G. J. (2017). Analysis of in-game communication as an indicator of recognition primed decision making in elite Australian rules football umpires. *Journal of Cognitive Engineering and Decision Making, 11*(1), 81–96.

Norman D. A. (1988). *The psychology of everyday things*. Basic Books, New York.

Norman, D. A. (2002). *The design of everyday things*. Basic Books, New York.

Norman D.A. & Draper S.W. (1986). *User Centered System Engineering*. Erlbaum.

Orphanides, A. K., & Nam, C. S. (2017). Touchscreen interfaces in context: A systematic review of research into touchscreens across settings, populations, and implementations. *Applied Ergonomics, 61,* 116–43.

Parasuraman, R., & Riley, V. (1997). Humans and automation: Use, misuse, disuse, abuse. *Human Factors, 39*(2), 230–253.

Parasuraman, R., & Wickens, C. D. (2008). Humans: Still vital after all these years of automation. *Human Factors*, *50*(3), 511–520.

Parnell, K. J., Stanton, N. A., & Plant, K. (2018). Where are we on driver distraction? Methods, approaches and recommendations. *Theoretical Issues in Ergonomics Science*, *19*(5), 578–605.

Parnell, K. J., Banks, V. A., Allison, C. K., Plant, K. L., Beecroft, P., & Stanton, N. A. (2019). Designing flight deck applications: Combining insight from end-users and ergonomists. *Cognition, Technology & Work*, *23*(2), 353–365.

Parnell, K. J., Banks, V. A., Plant, K. L., Griffin, T. G. C., Beecroft, P., & Stanton, N. A. (2021a). Predicting design induced error on the flight deck: An aircraft engine oil leak scenario. *Human Factors*. 36(6), 938–955.

Parnell, K. J., Wynne, R. A., Griffin, T. G., Plant, K. L., & Stanton, N. A. (2021b). Generating design requirements for flight deck applications: Applying the Perceptual Cycle Model to engine failures on take-off. *International Journal of Human–Computer Interaction*, 37(7), 611–629.

Parnell, K. J., Wynne, R. A., Plant, K. L., Banks, V. A., Griffin, T. G., & Stanton, N. A. (2022a). Pilot decision-making during a dual engine failure on take-off: Insights from three different decision-making models. *Human Factors and Ergonomics in Manufacturing & Service Industries*, 32(3), 268–285.

Parnell, K. J., Fischer, J. E., Clark, J. R., Bodenmann, A., Galvez Trigo, M. J., Brito, M. P., … & Ramchurn, S. D. (2022b). Trustworthy UAV relationships: Applying the Schema

Action World taxonomy to UAVs and UAV swarm operations. *International Journal of Human–Computer Interaction*, 1–17.

Parnell, K. J., Pope, K. A., Hart, S., Sturgess, E., Hayward, R., Leonard, P., & Madeira-Revell, K. (2022c). 'It's a man's world': a gender-equitable scoping review of gender, transportation, and work. *Ergonomics*, 1–17.

Parnell, K. J., Stanton, N. A., Banks, V. A., & Plant, K. L. (2023). Resilience engineering on the road: using operator event sequence diagrams and system failure analysis to enhance cyclist and vehicle interactions. *Applied Ergonomics*, *106*, 103870.

Plant, K. L., & Stanton, N. A. (2012). Why did the pilots shut down the wrong engine? Explaining errors in context using Schema Theory and the Perceptual Cycle Model. *Safety Science, 50*(2), 300–315.

Plant, K. L., & Stanton. N. A. (2013). The explanatory power of schema theory: Theoretical foundations and future applications in ergonomics. *Ergonomics, 51*(6), 1–15.

Plant, K. L., & Stanton, N. A. (2014). The process of processing: Exploring the validity of Neisser's perceptual cycle with accounts from critical decision-making in the cockpit. *Ergonomics, 58*, 909–923.

Plant, K. L., & Stanton, N. A. (2016). The development of the Schema World Action Research Method (SWARM) for the elicitation of perceptual cycle data. *Theoretical Issues in Ergonomics Science, 17*(4), 376–401.

Plattner, H. (2010). An introduction to design thinking: Process Guide. Available at https:web.stanford.edu. Accessed 28 Apr 22.

Rasmussen, J. (1974). The Human Data Processor as a System Component. Bits and Pieces of a Model. Riso-M-1722, Roskilde, Denmark: Risø National Laboratory.

Rasmussen, J. (1983). Skills, rules, and knowledge; signals, signs, and symbols, and other distinctions in human performance models. *IEEE transactions on Systems, Man, and Cybernetics, 3*, 257–266.

Read, G. J., Shorrock, S., Walker, G.H., & Salmon, P.M. (2021) State of science: evolving perspectives on 'human error', *Ergonomics, 64*(9), 1091–1114.

Reason, J. (1990). *Human Error*. Cambridge University Press.

Reeves, B., & Nass, C. (1996). *The media equation: How people treat computers, television, and new media like real people and places*. Stanford, CA: CSLI Publications

Revell, K. M., Richardson, J., Langdon, P., Bradley, M., Politis, I., Thompson, S., Skrypchuck, L., O'Donoghue, J., Mouzakitis, A., & Stanton, N. A. (2020). Breaking the cycle of frustration: Applying Neisser's Perceptual Cycle Model to drivers of semi-autonomous vehicles. *Applied Ergonomics, 85*, 103037.

RocketRoute. (2018). Take off from anywhere with RocketRoute HEMS and HELI mobile app. Available at: https://rocketroute.com. [Accessed 29 Jan 2019].

Rolfe, J. M., & Staples, K. J. (1988). *Flight simulation (No. 1)*. Cambridge University Press.

Roth, E. M., Sushereba, C., Militello, L. G., Diiulio, J., & Ernst, K. (2019). Function allocation considerations in the era of human autonomy teaming. *Journal of Cognitive Engineering and Decision Making, 13*(4), 199–220.

Rudd, J., Stern, K., & Isensee, S. (1996). Low vs. high-fidelity prototyping debate. *Interactions, 3*(1), 76–85.

Salas, E. (2004). Team methods. In N. A. Stanton, A. Hedge, K. Brookhuis, E. Salas, & H. Henrick, (Eds.) *Handbook of Human Factors Methods*, CRC Press

Salmon, P. M., & Read, G. J. (2019). Many model thinking in systems ergonomics: a case study in road safety. *Ergonomics, 62*(5), 612–628.

Salmon, P. M., Read, G. J. M., Walker, G. H., Lenné, M. G., & Stanton, N. A. (2019). *Distributed situation awareness in road transport. Theory, measurement, and application to intersection design*. CRC Press.

Salmon, P., Stanton, N., Walker, G., & Green, D. (2006). Situation awareness measurement: A review of applicability for C4i environments. *Applied Ergonomics, 37*(2), 225–238.

Salmon, P. M., Stanton, N. A., Walker, G. H., Jenkins, D., Ladva, D., Rafferty, L., & Young, M. (2009). Measuring Situation Awareness in complex systems: Comparison of measures study. *International Journal of Industrial Ergonomics, 39*(3), 490–500.

Salmon, P. M., Stanton, N. A., Young, M. S., Harris, D., Demagalski, J. M., Marshall, A., Waldman, T., & Dekker, S. (2002). Using existing HEI techniques to predict pilot error: A comparison of SHERPA, HAZOP and HEIST. In *Proceedings of the 2002 International Conference on Human-Computer Interaction in Aeronautics* (pp. 126–130).

Salmon, P. M., Walker, G. H., Read, G. J., Goode, N. & Stanton, N. A. (2017). Fitting methods to paradigms: Are ergonomics methods fit for systems thinking? *Ergonomics, 60* (2): 194–205.

Salmon, P. M., Walker, G. H., & Stanton, N. A. (2016). Pilot error versus sociotechnical systems failure: A distributed situation awareness analysis of Air France 447. *Theoretical Issues in Ergonomics Science, 17* (1): 64–79.

Samn, S. W., & Perelli, L. P. (1982). *Estimating aircrew fatigue: a technique with application to airlift operations(Final Report).* School of Aerospace Medicine Brooks AFB TX

Sarter, N. B., & Woods, D. D. (1995). How in the world did we ever get into that mode? Mode error and awareness in supervisory control. *Human Factors, 37*, 5–19.

Saunders, B., Sim, J., Kingstone, T., Baker, S., Waterfield, J., Bartlam, B., ... & Jinks, C. (2018). Saturation in qualitative research: exploring its conceptualization and operationalization. *Quality & Quantity, 52*(4), 1893–1907.

Shackel, B., 1986. Ergonomics in design for usability. In: M.D. Harrison and A.F. Monk, eds. *2nd conference of the British computer society human computer special interest group.* Cambridge: University Press, 44–64, 23–26 September 1986 York.

Shapiro, C. M., Flanigan, M., Fleming, J. A., Morehouse, R., Moscovitch, A., Plamondon, J., Reinish, L., & Devins, G. M. (2002). Development of an adjective checklist to measure five FACES of fatigue and sleepiness: data from a national survey of insomniacs. *Journal of Psychosomatic Research, 52*(6), 467–473.

Shneiderman, B. (1992). *Designing the user interface: Strategies for effective human-computer interaction.* Addison-Wesley, New York

Simon, H. A. (1955). A behavioral model of rational choice. *Quarterly Journal of Economics,* 69, 99–118.

Simpson, P. A. (2001). Naturalistic decision making in aviation environments. Defence Science and Technology Organisation Victoria (Australia). Air Operations Division, Aeronautical and Maritime Research Lab. DSTO-GD-0279.

Singer, G., & Dekker, S. (2001). The ergonomics of flight management systems: Fixing holes in the cockpit certification net. *Applied Ergonomics, 32*(3), 247–254.

Skaves, P. (2011). Electronic flight bag (EFB) policy and guidance. *Proceedings of IEEE/AIAA 30th Digital Avionics Systems Conference,* 16–20 Oct. 2011, Seattle, WA, USA.

Smith, K., & Hancock, P. A. (1995). Situation awareness is adaptive, externally directed consciousness. *Human Factors, 37*(1), 137–148.

Snyder, C. (2003). *Paper prototyping: The fast and easy way to design and refine user interfaces.* Morgan Kaufman.

Sorensen, L. J., Stanton, N. A., Banks, A. P. (2011). Back to SA school: Contrasting three approaches to situation awareness in the cockpit. *Theoretical Issues in Ergonomics Science, 12*(6), 451–471.

Sorensen, L. J., and Stanton, N. A., (2015). Exploring compatible and incompatible transactions in teams. *Cognition, Technology and Work, 17*(3), 367–380.

Sparaco, P. (1955). Airbus seeks to keep pilot, new technology in harmony. *Aviation Week and Space Technology*, January 30, 62–63

Stanton, N., (2004). Systematic Human Error Reduction and Prediction Approach (SHERPA). In: N. Stanton, A. Hedge, K. Brookhuis, E. Salas and H. Hendrick, ed., *Handbook of Human Factors and Ergonomic Methods* 1st ed. Boca Raton: CRC Press, pp.368–378.

Stanton, N. A., & Baber, C. (1992). Usability and EC directive 90270. *Displays, 13*(3), 151–160.

Stanton, N. A. & and Baber, C., (1996). A systems approach to human error identification. *Safety Science*, 22, 215–228.

Stanton, N. A., & Baber, C. (2002) Error by design: methods for predicting device usability. *Design Studies, 23*, 363–384.

Stanton, N. A., & Baber, C. (2008). Modelling of human alarm handling response times: A case study of the Ladbroke Grove rail accident in the UK. *Ergonomics, 51*(4), 423–440.

Stanton, N. A., Brown, J. W., Revell, K., Kim, J., Richardson, J., Langdon, P., Bradley, M., Caber, N., Skrypchuk, L., & Thompson, S. (2022a). OESDs in an on-road study of semi-automated vehicle to human driver handovers. *Cognition, Technology & Work, 24*(2), 317–332.

Stanton, N. A., Brown, J., Revell, K. M., Langdon, P., Bradley, M., Politis, I., Skrypchuk, L., Thompson., S., & Mouzakitis, A. (2022b). Validating Operator Event Sequence Diagrams: The case of an automated vehicle to human driver handovers. *Human Factors and Ergonomics in Manufacturing & Service Industries, 32*(1), 89–101.

Stanton, N. A. & Harvey, C., (2017). Beyond human error taxonomies in assessment of risk in sociotechnical systems: a new paradigm with the EAST 'broken-links' approach. *Ergonomics, 60*(2) 221–233.

Stanton, N. A., Harvey, C., Plant, K. L., & Bolton, L. (2013). To twist, roll, stroke or poke? A study of input devices for menu navigation in the cockpit. *Ergonomics, 56*(4), 590–611

Stanton, N. A., Li, W-C. & Harris, D., (2019a). Editorial: Ergonomics and human factors in aviation. *Ergonomics, 62*(2), 131–137.

Stanton, N. A., Plant, K. L., Revell, K. M. A., Griffin, T. G. C., Moffat, S. and Stanton, M. J., (2019b). Distributed cognition in aviation operations: A gate-to-gate study with implications for distributed crewing. *Ergonomics, 62*(2) 138–155.

Stanton, N. A., Rafferty, L. A., Salmon, P. M., Revell, K. M. A., McMaster, R., Caird-Daley, A. & Cooper-Chapman, C. (2010). Distributed decision making in multi-helicopter teams: case study of mission planning and execution from a non-combatant evacuation operation training scenario. *Journal of Cognitive Engineering and Decision Making, 4* (4), 328–353.

Stanton, N. A., Roberts, A. P. J, & Fay, D.T., (2017a). Up periscope: Understanding submarine command and control teamwork during a simulated return to periscope depth. *Cognition, Technology and Work, 19*(2–3): 399–417

Stanton, N. A., Salmon, P., Harris, D., Marshall, A., Demagalski, J., Young, M. S., Waldmann, T., & Dekker, S. (2009). Predicting pilot error: Testing a new methodology and a multi-methods and analysts approach. *Applied Ergonomics, 40*(3), 464–471.

Stanton, N. A., Salmon, P. M., Rafferty, L. A., Walker, G. H., Baber, C., & Jenkins, D. P. (2013). *Human factors methods: A practical guide for engineering and design*. Ashgate Publishing Ltd.

Stanton, N. A., Salmon, P. M., Walker, G. H., & Jenkins, D. (2009). Genotype and phenotype schemata and their role in distributed situation awareness in collaborative systems. *Theoretical Issues in Ergonomics Science, 10*(1), 43–68.

Stanton, N. A., Salmon, P. M., Walker, G. H., Salas, E., & Hancock, P. A. (2017). State-of-science: situation awareness in individuals, teams and systems. *Ergonomics, 60*(4), 449–466.

Stanton, N. A., & Stevenage, S. V. (1998). Learning to predict human error: issues of acceptability, reliability and validity. *Ergonomics, 41*(11), 1737–1756.

Stanton, N. A., Stewart, R., Harris, D., Houghton, R. J., Baber, C., McMaster, R., Salmon, P., Hoyle, G., Walker, G., Young M. S., Linsell, M., Dymott, R., & Green, D. (2006). Distributed situation awareness in dynamic systems: Theoretical development and application of an ergonomics methodology. *Ergonomics*, 49(12–13), 1288–311.

Stanton, N. A., & Walker, G. H. (2011). Exploring the psychological factors involved in the Ladbroke Grove rail accident. *Accident Analysis & Prevention, 43*(3), 1117–1127.

Stanton, N., & Young, M. (1998). Is utility in the mind of the beholder? A study of ergonomics methods. *Applied Ergonomics, 29*(1), 41–54.

Stanton, N. A., & Young, M. S. (1999). What price ergonomics? *Nature, 399*(6733), 197–198.

Stanton, N. A., & Young, M. S. (2003). Giving ergonomics away? The application of ergonomics methods by novices. *Applied Ergonomics, 34* (5), 479–490.

Stanton, N. A., Young, M. S. & Harvey, C. (2014) *A Guide to Methodology in Ergonomics: Designing for Human Use (second edition).* CRC Press: London, UK.

Stanton, N. A., Young, M. S., Salmon, P., Marshall, A., Waldman, T., & Dekker, S. (2002). Predicting pilot error: assessing the performance of SHERPA. Johnson, C.W. (Eds.), *21st European Annual Conference on Human Decision Making and Control,* Glasgow, Scotland. 47–51.

Swaminathan, S., & Smidts, C. (1999). The event sequence diagram framework for dynamic probabilistic risk assessment. *Reliability Engineering & System Safety, 63*(1), 73–90.

Szalma, J. L. (2014). On the application of motivation theory to human factors/ergonomics: Motivational design principles for human–technology interaction. *Human Factors, 56*(8), 1453–1471.

Taylor, R. (1990). Situational Awareness Rating Technique (SART): The development of a tool for aircrew systems design (AGARD-CP-478). In *Situational Awareness in Aerospace Operations* (pp. 3:1–17). NATO-AGARD: Neuilly Sur Seine, France.

Tromp, N., & Hekkert, P. (2016). Assessing methods for effect-driven design: Evaluation of a social design method. *Design Studies, 43,* 24–47.

Turiak, M., Novák-Sedláčková, A., & Novák, A. (2014). Portable electronic devices on board of airplanes and their safety impact. In Mikulski, J. (Ed.), Communications in Computer and Information Science, 471, 29–37.

Van Der Laan, J. D., Heino, A., & De Waard, D. (1997). A simple procedure for the assessment of acceptance of advanced transport telematics. *Transportation Research Part C: Emerging Technologies, 5*(1), 1–10.

Vidulich, M. A., & Hughes, E. R. (1991). Testing a subjective metric of situation awareness. In *Proceedings of the Human Factors Society Annual Meeting*, 35, (18), pp. 1307–1311. September. Sage CA: Los Angeles, CA: SAGE Publications.

Vidulich, M. A., Wickens, C. D., Tsang, P. S., & Flach, J. M. (2010). Information processing in aviation. In *Human Factors in Aviation* (pp. 175–215). Academic Press. Chicago.

Virzi, R. A. (1992). Refining the test phase of usability evaluation: How many subjects is enough? *Human Factors, 34*(4), 457–468.

Virzi, R. A. (1997). Usability inspection methods. In (eds) M.G. Helander, T.K. Landauer and P.V. Prabhu. *Handbook of human-computer interaction* (pp. 705–715). North-Holland.

Waag, W. L., & Houck, M. R. (1994). Tools for assessing situational awareness in an operational fighter environment. *Aviation, Space, and Environmental Medicine, 65*(5, Sect 2, Suppl), A13–A19.

Walker, G. H., Salmon, P. M., Bedinger, M. & Stanton, N. A., (2017). Quantum ergonomics: shifting the paradigm of the systems agenda. *Ergonomics, 60*(2) 157–166.

Walters, A. (2002). *Crew resource management is no accident.* Aries, Wallingford.

Wickens, C. D. (2002). Multiple resources and performance prediction. *Theoretical Issues in Ergonomics Science, 3*, 159–177.

Williams, P. D. (2017). Increased light, moderate, and severe clear-air turbulence in response to climate change. *Advances in Atmospheric Sciences, 34*(5), 576–86.

Winter, S. R., Milner, M. N. Rice, S., Bush, D., Marte, D. A., Adkins, E., Roccasecca, A., Rosser, T. G., & Tamilselvan, G. (2018). Pilot performance comparison between electronic and paper instrument approach charts. *Safety Science, 103*, 280–286.

Woods, D. D., & Watts, J. C. (1997). How not to have to navigate through too many displays. In *Handbook of Human-Computer Interaction* (Second Edition), pp. 617–650.

Wright Brothers Aerospace Company (2010) *Wright Flyer*. Retrieved from www.wright-broth ers.org/.

Wynne, R. A., Beanland, V., & Salmon, P. M. (2019). Systematic review of driving simulator validation studies. *Safety Science, 117*, 138–151.

Yang, Q., Sun, X., Liu, X., & Wang, J. (2020). Multi-agent simulation of individuals' escape in the urban rainstorm context based on dynamic recognition-primed decision model. Water, 12(4), 1190.

Young, M. S., & Stanton, N. A. (2002). Attention and automation: New perspectives on mental underload and performance. *Theoretical Issues in Ergonomics Science, 3*(2), 178–194.

Index

A

acceptance (user) 22–4, 38
advancing current systems 59
attention 26, 27, 47

B

between measures analysis 14

C

comfort 52
communication 17, 27, 30
confidence (user) 23–4
counterbalancing 14
Critical Decision Making (CDM) method 38, 106

D

decision ladders 57, 104, 108–11
decision making 16, 26, 28, 32, 38, 57, 104
decision modelling 16
decision aid 104–18
design 23, 25, 27, 29, 39, 46, 48, 52–8, 149
design (methods) 31
design (principles) 43–5, 49
design (process) 16, 18, 21, 31, 35, 38, 52–8, 152
design (research) 19
design (requirements) 24, 25, 27, 46–8, 55
design (recommendations) 52, 76
design (standards) 19
design (user-centred) 17–18, 22, 39, 80
design (usability principles) 43–5
design (workshops) 16
design with intent (DWI) 31, 34, 35, 39, 55, 79, 80, 89, 91–2, 119, 128

E

effectiveness 23, 25, 29, 41, 42, 47–9, 52
ergonomics (principles) 23, 31, 41
error 23, 25, 30, 34, 43, 44, 49, 52, 66
error (modelling) 56
error (human) 25, 33, 45, 46, 55
evaluation 17, 21, 29, 34, 36, 41, 42, 45, 48, 50, 53, 57, 58, 151
evaluation (context of use) 36
evaluation (simulator) 36, 37, 39
evaluation (wire-frame) 36, 117–19
expertise 14, 16, 18, 21, 31, 37, 55

F

fan damage 105–14
fatigue 26, 27, 42, 50
fidelity 11, 16, 36
flight deck 26–8, 30, 36, 37, 39, 41, 46, 48, 50, 52
flight management system 46, 60, 61, 97–9
focus groups 31

G

generating requirements 53, 55

H

heuristic evaluation 36, 57, 116–26, 151
Hierarchical Task Analysis (HTA) 29, 30, 33, 39, 61, 68
human error identification (HEI) 25, 29, 30, 33, 86
human error template (HET) 30
human-machine interface (HMI) 39, 43
hypothesis 11–13

I

interviews 16, 17, 22, 27–8, 32, 38, 55

L

layout analysis 31
link analysis 31

M

methods 17–20, 22–3, 26, 30, 56
methods (generating requirements) 27, 55
methods (objective) 22
methods (physiological) 26
methods (subjective) 15, 23, 24
model 56–7, 150
modelling 52
modelling (engineering integration) 95–104
modelling (user-behaviour) 104
motivation 12, 24, 42

N

NASA-TLX 15, 52, 136
naturalistic decision making 57, 104–19

O

oil starvation 68–76, 80–5, 124–31

Printed in the United States
by Baker & Taylor Publisher Services